William Charles Arlington Blew

Light Horses

Breeds and Management

William Charles Arlington Blew

Light Horses
Breeds and Management

ISBN/EAN: 9783744648226

Printed in Europe, USA, Canada, Australia, Japan

Cover: Foto ©berggeist007 / pixelio.de

More available books at **www.hansebooks.com**

LIVE STOCK HANDBOOKS.

Edited by JAMES SINCLAIR, *Editor of* "*Live Stock Journal,*" "*Agricultural Gazette,*" *&c.*

No. II.

LIGHT HORSES.

BREEDS AND MANAGEMENT

BY

W. C. A. BLEW, M.A.; WILLIAM SCARTH DIXON;
DR. GEORGE FLEMING, C.B., F.R.C.V.S.;
VERO SHAW, B.A.; ETC.

THIRD EDITION.

ILLUSTRATED.

London:
VINTON & COMPANY, LTD.,
9, NEW BRIDGE STREET, LUDGATE CIRCUS, E.C.

1898

CONTENTS.

	PAGE
CHAPTER I.—The Thoroughbred Horse	1
CHAPTER II.—The Hackney Horse	23
CHAPTER III.—Cleveland Bays and Yorkshire Coach Horses	51
CHAPTER IV.—The Arabian Horse	81
CHAPTER V.—The American Trotting Horse	94
CHAPTER VI.—The Hunter	105
CHAPTER VII.—The Hack	126
CHAPTER VIII.—Ponies	136
CHAPTER IX.—Asses and Mules	153
CHAPTER X.—Management of Light Horses	159
CHAPTER XI.—Diseases and Injuries to which Light Horses are Liable	188

ILLUSTRATIONS.

		PAGE
Thoroughbred Horse, Orme	...To Face	6
Thoroughbred Mare, Plaisanterie...	,,	11
Thoroughbred Stallions, King Herod and Flying Childers	,,	17
Thoroughbred Stallion, Ruddigore	,,	21
Hackney Stallion, Danegelt	,,	27
Hackney Stallion, M.P....	,,	31
Hackney Stallion, General Gordon	,,	37
Hackney Mare, Lady Wilton	,,	43
Hackney Mares	,,	47
Cleveland Bay Stallion, Master Frederick ...	,,	69
Cleveland Bay Stallion, Sultan	,,	63
Yorkshire Coaching Stallion, Prince of Wales	,,	77
Arabian Stallion, Kahalet	,,	87
Anglo-Arabian, Khaled ...	,,	91
Arabian Stallion, Speed of Thought	,,	93
Thoroughbred Stallion Mambrino...	,,	95
American Trotter	,,	103
Group of Hunters	,,	109
Hunter Mare ...	,,	111
Hunter	,,	115
Hunter Sire ...	,,	117
Hack Hunter ...	,,	129
Carriage Horses	,,	133
Tandem Team...	,,	135
Pony Mares ...	,,	144
Pony Stallion, Sir George	,,	149
Shetland Pony, Good Friday	,,	151
Welsh Pony, Tommy ...	,,	152

LIGHT HORSES.
BREEDS AND MANAGEMENT.

CHAPTER I.

THE THOROUGHBRED HORSE.

It is by no means easy to frame a succinct definition of the thoroughbred horse. We know that no horse is accepted as thoroughbred unless he appears duly registered in the Stud Book, and so, to save trouble, we may take this as the criterion. The blood horse, however, like the fox-hound, is after all an animal of composite breed; that is to say, time was when he did not exist; and no horse presenting the features of the modern English thoroughbred was, at one period, to be found in England.

To show this conclusively, would be to write in detail the history of the English horse—an unnecessary task, inasmuch as this subject has already been fully dealt with in many books. It will, therefore, be sufficient for our present purpose, if we take the time of King Charles II. as an important landmark, and briefly trace the history of the English horse up to that reign, before entering into any sort of disquisition upon what we now call the thoroughbred horse.

We need spend no time in enquiring what sort of horses they were which so excited the admiration of Julius Cæsar;

but they were good enough to induce the Conqueror to take a good many away with him. When the Romans established themselves in Britain, they found it expedient to send over a large body of cavalry to assist in maintaining order, and the horses of these soldiers were doubtless crossed with the native stock; and so the British horse, no matter what it was like, received its first modification or cross. Whether this cross improved it or not, is not to the purpose; we merely note the fact that it was, so far as known, the first step towards manufacturing what we venture to describe as a composite breed, more especially as what we have generically termed Roman horses were collected, not only from Italy, but from Gaul and Spain.

Then, again, it has been said that the time of the Roman occupation of Britain, saw the first importation of Eastern blood, as Severus is reported to have raced *bona fide* Arabs at Wetherby, Yorkshire. This story, however, lacks verification, and may be passed over without any importance being attached to it. Rather more trustworthy, perhaps, is the statement that Hugh Capet, of France, while courting Ethelwitha, sister of King Athelstan, sent the latter a present of some German "running horses," partly, no doubt, in honour of the event which was soon to take place; and partly by way of congratulation to the King on his having subdued the rebellious portions of the Heptarchy. At any rate, whatever may have been the motive, we are perhaps justified in concluding that our native horses were crossed with these new comers. This, it may be noticed, is the first mention of "running horses," and the conclusion may be drawn that they were of a lighter build than our own native steeds, though we are still left in the dark as to what sort of an animal the German running horse was. Here, however, was the introduction of another strain of blood.

William the Conqueror's own charger is said to have been a Spanish horse, and Roger de Boulogne, Earl of Shrewsbury, is reported to have introduced Spanish horses on his estates.

The Conquest, therefore, brought with it sundry strains of foreign blood, which must necessarily have had its influence in more or less changing the appearance of our native horses.

As Eastern blood has admittedly been so potent a factor in making our English thoroughbred the horse he is, we may just pause to point out that in the reign of Henry I., we come across the first recorded importation of an Eastern horse. The story goes that Alexander I., King of Scotland, presented to the Church of St. Andrew (*inter alia*) an Arabian horse. Mention has already been made of the Arabs Severus is said to have raced in Yorkshire, but we may search in vain for any hint as to how they got here.

That an opportunity was missed at the time of the Crusades is tolerably certain; and, if we make passing notice of the fact that Richard Cœur de Lion is reported to have possessed two horses he purchased from Cyprus, and which were probably of Eastern origin, we may go on to the reign of Edward III., for most of King John's exertions were so extended upon war and heavy horses, though at the same time he did not neglect the race course, as he imported many Eastern horses. Edward III., however, bought fifty Spanish horses, believing that their blood would materially improve the native breed, but he is said to have almost repented of his extravagance on finding that they had cost him no less than £13 6s. 8d. per head. This King, who was unquestionably a sportsman in his way, had running horses, and was fully alive to the importance of trying to get a lighter and faster horse than the ponderous animals which were required to carry the armoured soldier in battle.

One of Henry VIII.'s officials was styled Master of the Barbary horses; but whether this very arbitrary monarch had any Barbs, or whether this was merely a generic term for his race horses is, we venture to think, doubtful; we do know, however, that he imported horses from Turkey, Spain, and Naples; while the Marquis of Mantua gave him some high-class mares, and the Duke of Arbino presented him with a stallion. Still, partly perhaps owing to the restrictions placed by the King upon breeding operations, we are unable to gather from the records that the stamp of horses improved to any great extent in his reign. With Queen Elizabeth's accession to the throne, however, a better state of things commenced. It is true that racing fell off; but, as a compensation, the breed was maintained to a great extent through Barbs, and Spanish horses descended from Barbs which were found on the ships captured by Lord Howard, of Effingham, when he routed the Armada.

Although James I. has often been sneered at on account of the manner in which he occasionally followed hunting and racing—in some ways he may remind us of Colonel Thornton—he was beyond doubt a sportsman somewhat in advance of his time. To confine ourselves, however, to the introduction of foreign strains of blood, it seems that a good many foreign horses were sent as presents to the English Court; half-a-dozen Barbs are said to have been brought to England by Sir Thomas Edmonds, who, as ambassador and traveller, had many opportunities of seeing Eastern sires, and who no doubt imported others of which we know nothing.

One imported horse, however, must be specially noticed—the Markham Arabian. So far as we can judge, this seems to have been a private purchase of the King's, prompted solely by his own desire to try an experiment. Possibly he may have remembered the Arab said to have been presented to the Church of St. Andrew about five hundred years before; and may have desired to try once more the effect of this blood. To put the matter shortly, the Markham Arabian appears to have been a failure. He was put into training, but could win no races; nor could any of his stock run. Prior to this time, there were as we have shown, a great many Eastern horses of one kind and another imported; but this Markham Arabian is the only one concerning which we have any details; and these might probably not have been forthcoming had it not been that the Duke of Newcastle saw him, thought him "small potatoes and few in a hill," and spoke of him in the most disparaging terms.

During this time, however, there were in England horses which could hold their own against all foreign importations. For this statement Gervase Markham is our chief authority. He may possibly have been imbued with a certain amount of patriotic admiration for home products, but this is what he wrote:—" Again, for swiftness, what nation hath brought forth that horse which hath exceeded the English ?—when the best Barbaries that ever were were in their prime, I saw them overrunne by a black hobbie at Salisbury, yet that hobbie was more overrunne by a horse called Valentine, which Valentine neither in hunting nor running was ever equalled, yet was a plain bred English horse both by syre and dam."

It is interesting, too, to note the description of the English horse as given by the same expert. "Some former writers," he says, "whether out of want of experience, or to flatter novelties, have concluded that the English horse is a great strong jade, deep ribbed, sid-bellied, with strong legges and good hoofes, yet fitter for the cart than either saddle or any working employment. How false this is all English horsemen knowe. The true English horse, him I meane that is bred under a good clime on firme ground, in a pure temperature, is of tall stature and large proportions; his head, though not so fine as either the Barberie's or the Turke's, yet is lean, long, and well fashioned; his crest is hie, only subject to thickness if he be stoned, but if he be gelded[*] then it is firm and strong; his chyne is straight and broad, and all his limbs are large, leane, flat, and excellently jointed."

A horse with a lean head, a good chine, and flat legs has certainly the attributes of a good one. As already mentioned, the Markham Arabian was invariably beaten on a race course; and it is important to bear all these matters in mind now that

---

[*] The practice of castrating horses is said to have been first practised in the time of Henry VII.

we are approaching a period at which the modern thoroughbred may be said to have been invented.

Continuing our notice of imported Eastern horses, we find that in the reign of Charles I., Sir Edward Harwood laments the scarcity of strong horses, giving as the reason that light and swift horses were bred for the purpose of racing; and, though Sir Edward may have been in error in supposing that strong horses fit for the cavalry soldier were scarce, his testimony to the existence of race horses helps us to understand that light horses were being bred with considerable care.

In this reign (Charles I.) the Duke of Buckingham brought over to England a horse known as the Buckingham Turk, which, from being sold to Mr. Helmsley, acquired the better known name of the Helmsley Turk. He does not appear to have been raced; but, as will hereafter appear, to have left his name in the Stud Book. The sad events in this country, which culminated with the execution of the King, not unnaturally retarded the breeding of horses; but Cromwell was clear enough to perceive that the country at large would benefit by the public attention being drawn to the importance of having suitable light horses for the cavalry; and so he kept his own stud and race horses; and, having at his elbow a Mr. Place, by reputation a skilful breeder and his stud master, he became possessed of the White Turk, imported by Mr. Place, in connection with whose descendants we shall have something to say presently.

Without going at undue and unnecessary length into the history of the English horse—and the history has been written by many pens—an attempt has been made to show, in the first place, how the native horses were probably altered in type, by being crossed with the different foreign horses which have from time to time been brought over to England; and, secondly, that there appears to have been in England a breed of horses, to wit, those described by Gervase Markham, which were, at any rate fast enough to beat the Eastern horses pitted against them.

THOROUGHBRED HORSE, ORME.
Winner of the Middlepark Plate, Eclipse Stakes, &c.
Bred by and the Property of the Duke of Westminster, K.G.

How these "hobbies" were bred we have no means of knowing, but many of them are said to have come from Ireland, and this is a rather curious circumstance in horse-breeding. Until the hairy-heeled cart horse was introduced into Ireland, the cart horse of the country was a clean-legged one, and it was from these that the famous Irish hunters came—hunters up to weight, and by no means lacking pace. Is it not, therefore, very probable that the race horse of former days may have been bred on similar lines?

What has been written above has gone to show that up to the time of the Commonwealth a good many external strains of blood had been grafted on to the native stock; even at this time the lightest and swiftest horse was a composite animal, more like our hunters—he could not have been bred to type; nor could he have shown the mark of any particular breed like the blood horse of to-day. Like our weight-carrying hunter, he must have been more or less a chance-bred animal, and in a kingdom of the blind where the one-eyed were kings, the fastest stood out from the rest of their composite bred brethren.

How far pedigrees were kept generally we have little means of knowing. But when all domestic matters were turned upside down by that disastrous upheaval which put a stop to everything except ill-feeling, it is more than probable that many of the records which had unquestionably been kept during the reign of James I. were destroyed, just as many ecclesiastical records were destroyed at the Reformation. At this stage, at all events, we are justified in arriving at the conclusion that there were in England different kinds of horses, and that from time to time the native stock had been crossed with various foreign strains; and in this state matters stood at the Restoration.

King Charles II. may not have been in all respects an ideal monarch; but it is to him that we owe the foundation of our present race of thoroughbred horses. It seems, however, to have been sometimes assumed that the thread of horse-breeding was cut when Charles II. came to the throne; and that what subsequently took place had no connection with what had gone before. In other words, the opinion seems to have been entertained that the stock which was in existence when Charles II. succeeded had nothing to do with the origin of the thoroughbred horse. To this doctrine we cannot subscribe, as there is no ground for thinking that the previously existing race horses were entirely tabooed by the "Merry Monarch;" in fact, there is every reason to come to the opposite conclusion.

Though James I. and Charles I. introduced Eastern blood, we do not read that they or any of their predecessors imported many Eastern mares; though, of course, it is only reasonable to suppose that a certain number did come to England. King Charles II., however, commissioned his Master of the Horse, Sir John Fenwick, who was also a breeder on his own account, to go abroad and bring back some mares of the best blood he could find. This was no doubt a congenial task to Sir John, who was a well-known racing man of that time; but whither he went, how many mares he bought, and what they were, are matters which cannot be stated with any certainty.

According to some authorities Sir John Fenwick went to Tangier and returned with four Barb mares; others say that eight were procured, while others again incline to the opinion that the new purchases included an Arab mare, and possibly a Turk. Be this as it may, these mares were known as Royal mares; but, inasmuch as some of their female offspring were also called Royal mares the exact number of the original stock cannot be ascertained by reference to any existing records.

It is in the time of King Charles II. that the modern thoroughbred horse may be said to have been invented. Sir John Fenwick brought back with him not only the Royal mares, but some sires as well; and from this period a constant stream of Eastern blood flowed into England.

So far as can be made out we know very little of what success attended the importation of the Royal mares. One of the Royal mares, however, was heavy in foal when she reached England, and soon after joining the King's stud, gave birth to the colt, afterwards known as Dodsworth, who was, of course, a pure Barb. Then, as now, private enterprise was largely expended upon horse-breeding; and several of the King's subjects, Lord Cullen, and Lord Conway, for example, imported several Eastern horses. On the death of Charles II. the Royal mares and other horses appear to have been sold. Dodsworth's dam, bought by Mr. Child from the stud master for 40 guineas, and several of the others apparently passed into the possession of Mr. John D'Arcy who seems to have founded a breeding stud at Sedbury soon after the Restoration. This, at any rate, we may infer from the fact that we find the names of the D'Arcy Grey Royal Mare, and three other D'Arcy Royal Mares. Lord D'Arcy also imported two Eastern sires, known respectively as D'Arcy's White Turk and D'Arcy's Yellow Turk, and these, by being put to the Royal, and other good mares, left their mark upon the earlier stock of this country.

We gather from the Stud Book that about 176 Eastern sires were imported from the time of King James I.; and of these 24 found their way to England in the reign of Queen Anne.

We would here again remind the reader that the thoroughbred race horse was not for the first time created by the importation of Eastern blood, though it unquestionably improved our native horses; but, amid the many importations which began in the time of the Stuarts, we lose sight of the old English race horse as he existed before the Restoration.

It would be mere plagiarism to discuss in detail all the Eastern horses the names of which are to be found; it will, therefore, suffice for this description of the manufacture of the blood-horse to make short mention of three sires who may be said to have accomplished great things for the breed of race horses, though it is incorrect to say that they, between them, made the thoroughbred.

The first is the Byerly Turk, so called because he was

ridden as a charger by Captain Byerly during King William's campaign in Ireland. When this horse came to England is not quite clear; but it was probably about the year 1689. Then, during the first dozen years of the 18th century, Mr. Darley, a Yorkshire gentleman, introduced the Darley Arabian (the sire of Childers), and, lastly, came the horse known as the Godolphin Arabian; but which was probably a Barb. He must have come to England about 1728; in 1730 and 1731 he was teaser to Hobgoblin, and might have been unknown to fame had it not so happened that on the refusal of Hobgoblin to serve Roxana, the Godolphin Arabian had to take his place; and the result was Lath, the first of his get.

It will, therefore, be seen from the dates of these horses that races had been held very long before their time; that Eastern horses had run, and been beaten by English horses; and that Eastern blood had been used. Consequently, it is evident that this illustrious trio of sires founded no new breed; they would only be crossed with the then living mares. And what were these mares? The female lines of race horses are too often neglected by writers; but if the matter be examined closely it will be found that there is a good deal of blood in the modern thoroughbred which is not of Eastern origin.

This point is clearly and forcibly put by that high authority, Mr. Joseph Osborne ("Beacon"), in the valuable introduction to the "Breeder's Handbook." In protesting against the oft-made assertion that the English thoroughbred owes his origin to the Byerly Turk, the Darley Arabian, and the Godolphin horse, Mr. Osborne says: "But in the female line there are undeniable proofs of important influence outside and anterior to the known Eastern sires; and here I maintain that, in considering the *origin* of the 'thoroughbred' as distinct from his Stud Book genealogy, it is most unreasonable—nay, even preposterous—to set aside the female sources. Indeed, if the investigation be pursued logically, that side must be the more interesting in this instance

because, the sires being admittedly Eastern, it becomes imperative to trace, if possible, the blood of their mates. If a potent average of the dams at the roots is found not to be Eastern, then it becomes obvious that any restrictive claim for purely Arab descent has no authentic basis in fact."

We may find much proof of what is advanced by Mr. Osborne by referring to the pedigree of the famous Eclipse. Marske was the sire of Eclipse, and in Marske's pedigree we find that Snake was by the Lister Turk, out of a mare by Hautboy; but the name of the mare which threw Snake's dam to Hautboy is not to be found. This is of itself presumptive, though not conclusive, evidence that the mare was English bred. Had she been an Eastern matron her identity would certainly have been established. Then, again, Grey Wilkes was by Hautboy, out of Miss D'Arcy's Pet Mare; but who was the sire of this Pet Mare cannot be ascertained. The inference is that the sire was an English horse; and this is all the more probable because Lord D'Arcy, as already pointed out, was one of the foremost breeders of the day; and as he mentions all his Eastern horses, he would certainly have kept record of this mare had he known her lineage. Of Coneyskins we have no knowledge beyond the fact that he was a son of the Lister Turk; his dam was probably an English mare; while we may search in vain for the breeding of the Old Clubfoot mare, except that she was by Hautboy; and yet she was the property of Mr. Crofts, who bred largely, as the term was understood in those days. Not to labour the matter out to an undue length, it may be shortly stated that there are flaws in the pedigree of Bay Bolton's sire; and it is curious to note that the identity of so many mares which were sent to Hautboy has been lost. Grey Hautboy, sire of Bay Bolton, was by Hautboy.

In the pedigree of Spiletta, the dam of Eclipse, we find sundry other blanks which cannot be filled in; and in each case the probability is that English blood should claim the honour of a place. Mr. Osborne says: " The thirteen unknown sources affect nine of the sixteen divisions, leaving the natural inference that the amount of English blood in the pedigree of Eclipse is almost as large as that of the Eastern sires; and it is impossible to tell the relative influence of either blood in the descent. . . . . The Eastern blood is unaffected, so far as Bartlett's Childers (son of the Darley Arabian), and no further; for his son, Squirt, inherits the unknown blood in two distinct lines from his dam, Sister to Old Country Wench; whilst Marske, the son of Squirt (and sire of Eclipse), has a far greater admixture of the unknown (but, as I assume, English) element through his dam, the daughter of Blacklegs, who has no fewer than seven blanks, or, in other words, only one of eight lines of descent can be traced to a purely Eastern source. What equitable claim, therefore, can be made to a purely Eastern descent on his sire's side, if both his sire and grandsire inherit so many strains to which no Eastern origin can be assigned?"

Mr. Osborne continues: "The origin of Eclipse traced on the side of his dam, Spiletta, is even more convincing as to the extravagant conclusions which have been made. Even the best influence of the Godolphin has commenced with the unknown element in his son Regulus, whose dam, Grey Robinson, is, of course, affected by the remarks above, concerning the Sister to Old Country Wench; whilst Mother Western, the maternal granddam of Eclipse, is conspicuously wanting in Eastern credit, since nothing is known of the dam of her sire, besides the discrepancies in Snake, and the 'unknown quantity' in her dam, the Old Montague mare, through the maternal descent of Merlin. I need say no more about this great pedigree. The evidences which have influenced my own judgment are before the reader in a way that enables him to form his own judgment independently. There is nothing revolutionary in what I have stated. The best authorities have referred, though only *en passant*, to the Eastern sires as *improvers;* but they have left the assumption that the old English influence was at once obliterated by them; and that to them alone is due the credit of the whole development."

Eclipse was foaled in 1764, and the above-quoted remarks show that there was in his veins a considerable amount of, to say the least, unknown blood; and as Pot-8-os, Waxy, Whalebone, Camel, Touchstone, Orlando, and Teddington were amongst his direct descendants, it follows that what Mr. Osborne designates the "unknown quantity," existed in those famous horses.

What has been said above justifies, we venture to think, the statement previously made—that the race horse of to-day is a composite animal; while it is not to be denied that the admixture of Eastern blood materially benefited our native stock. Then the time came when Eastern sires were no longer used to develop the thoroughbred; and the volumes of the Stud Book now tell their own tale.

In order to give the reader who may not care for deep research into the Stud Book, a general idea of the families and roots, we may just run quickly through some of the lines. Of the Helmsley Turk, we have already made mention, and it is only necessary to say here, that it is as the sire of Bustler his memory has chiefly been held in veneration, as Bustler's blood is of importance in the older pedigrees. Here, again, Place's White Turk is entitled to honourable mention, as he not only sired some good racehorses, but at the stud helped breeders with some of his female descendants, while the strain has been handed down to us through Matchem and Woodpecker.

As already pointed out, the three sires, the Byerly Turk, the Darley Arabian, and the Godolphin Arab or Barb did not found a new breed: they were merely fresh infusions. They are, however, commonly spoken of as the fountain heads from which our best known horses have mainly sprung, the Byerly Turk through Herod, the Darley Arabian through Eclipse, and the Godolphin horse through Matchem. If the male lines alone be considered, the modern racehorse has more of the

Eclipse blood in him than he has of that of either Herod or Matchem, for from Eclipse have proceeded very many families. In the same way it can be shown that, with reference to the male line alone, Herod is next represented, and Matchem least of all. When, however, we come to take the female lines into consideration as well, the complexion of the case is somewhat changed, and it will be found that, with scarcely an exception, the foremost horses of to-day have more of the blood of Herod in their composition than of Eclipse, that is to say, they represent the Byerly Turk to a greater extent than they do the Darley Arabian.

In a most learned and carefully thought out article on "The Blood of our Thoroughbred Horses," which appeared in the *Field* for the 8th and 29th of May, 1886, and in the *Rural Almanack* for 1887, all this and much more is cleverly shown by carefully prepared tables, and the writer there remarks "the representation of the Godolphin line of descent (Matchem) is 'nowhere' in comparison with the representation of the Darley line (Eclipse) as regards the number of direct male descendants; but the descendants, in respect to proportion of blood, do not derive from the Darley Arabian one half as much as they do from the Godolphin. And it is even more marked in the line of Herod, for in that sire there was not one drop of Godolphin blood, whereas Herod's descendants in the present day derive from the Godolphin three times as much as from the Byerly Turk which they are considered to represent." The reader, therefore, will understand that, although certain horses are described as being descendants of one of the three Eastern sires, or of Herod, Eclipse, or Matchem, this does not mean that they do not include much blood of the others, for, as time has gone on, the three strains have become commingled.

As the Byerly Turk came to England before either the Darley Arabian or the Godolphin, we will speak of him first. The famous horse, Herod, was great grandson of Jigg, who was son of the Byerly Turk; and until Partner, one of his sons, was six years old, Jigg was merely a travelling stallion in Lincolnshire. Among Herod's best sons were Highflyer Woodpecker, Florizel and Phenomenon. From this strain we have Highflyer, Sir Peter Teazle (commonly called Sir Peter), Selim, Pantaloon, Partisan, Bay Middleton, Gladiator, Glaucus, Sweetmeat, The Flying Dutchman and Wild Dayrell. From two mares by the Byerly Turk there are descended in the female line, as the new edition of the first volume of the Stud Book tells us, Bend Or, Robert the Devil, Uncas, Speculum, Blue Gown, Craig Millar, Paradox, The Bard, Minting, &c.

From the Darley Arabian was descended Eclipse, and also King Fergus, Pot-8-os, Waxy, Whalebone, Orville, Whisker, Sir Hercules, Touchstone, Irish Bird-catcher, Lanercost, Blacklock, the Baron, Rataplan, Stockwell, King Tom, and Newminster, whose son was the famous Hermit.

Lastly comes the Godolphin Arabian, said to be really a Barb, and in some ways the most important of these three Eastern sires. From him we get Matchem, Conductor, Humphrey Clinker, Melbourne, West Australian, Prime Minister, Knight of the Garter, Knight of Kars, &c.

Such, then, is a necessarily imperfect outline sketch of the manner in which the modern thoroughbred has been built up or manufactured. In the course of his development there is one fact which everyone admits—he has increased in size; and having said this, we have said everything on which men are universally agreed. Some people say that the modern race horse is not as stout as he was. Of this opinion is Mr. Joseph Osborne. In the "Horse Breeders' Handbook" he says that the Irish-bred horses, Byron, Paladour and Napoleon "found it no trouble to run four miles under eight minutes," while on a subsequent page he quotes the Earl of Stradbroke as follows:— "My firm belief is that there are not four horses in England now that could run over the Beacon Course (4 miles 1 furlong 135 yards) at Newmarket within eight minutes, which in my younger days I used to see constantly done."

Of course, in making these comparisons much will depend upon the credit attaching to the old records. About 1721 Flying or Devonshire Childers is said to have run a trial over the Round Course (3 miles, 4 furlongs, 93 yards) at Newmarket, in 6 mins. 41 secs., that is to say, at an average rate of speed of 1 min. 50 secs. for a mile, though he was carrying 9 stone 2lbs. On a subsequent occasion Childers is said, when carrying the same weight, to have run the Beacon Course (4 miles, 1 furlong, 138 yards) in 7 mins. 30 secs. In this instance we find that each mile took Childers less time to cover than when he was running the shorter course, for the average over the four miles is no more than 1 min. 47 secs. and a fraction of a second for each mile.

Matchem is credited with having even surpassed this performance, the report being that he ran the Beacon Course in 7 mins. 20 secs., or a mile in 1 min. 44 secs.; but then his weight was 8 stone 7lbs. instead of 9 stone 2lbs.

Now if these times be compared with those of more modern horses over shorter courses, it will be to the disadvantage of the moderns, as the following figures will show, and a few of the best times have been selected as given in *Ruff*:—

In 1846, the first year in which the time is given in *Ruff*, Pyrrhus I. won the Derby in 2 min. 55 secs.; and if we take the Derby Course at a mile and a half, Pyrrhus I. ran at the rate of a mile in 1 min. 56⅔ secs. The average of the Flying Dutchman (1849), Daniel O'Rourke (1852), and Ellington (1856), in still slower, as they took 3min.; 3 min. 2 secs.; and 3 min. 4 secs. respectively to get over the mile and a-half. Kettledrum, Blair Athol, Merry Hampton, and Ayrshire show better time, as they covered the distance in 2 min. 43 secs., that is to say, at an average rate of speed of 1 min. 48⅔ secs. for a mile.

The St. Leger Course is given in *Ruff* as 1 mile, 6 furlongs, and 132 yards; and in 1888 Seabreeze's time, the fastest on record, was 3 min. 11⅖ secs., which gives an average of 1 min. 45½ secs.

From a Painting by J. N. Sartorius.
KING HEROD (Foaled 1758).
FLYING CHILDERS (Foaled 1715).

Then if we take White Feather's time for the Goodwood Stakes of 1891, we find the time for a mile is 2 mins. 3$\frac{1}{20}$ secs., the distance being set down as two miles and a-half. Put in tabular form the averages compare thus :—

	m. f. yds. 4 1 138	m. f. yds. 3 4 93	m. f. yds. 2 4 0	m. f. yds. 1 6 132	m. f. yds. 1 4 0
Childers	1 min. 47 secs.	1 min. 50 secs.	—	—	1 min. 56¾ secs.
Matchem	1 min. 44 secs.	—	—	—	1 min. 48 secs.
Pyrrhus I.	—	—	—	—	—
Kettledrum	—	—	—	—	—
Seabreeze	—	—	—	1 min. 45½ secs.	—
White Feather	—	—	2 min. 3$\frac{1}{20}$ sec.	—	—

The distance at the top of each column denotes the whole distance run, and the figures in the columns the average pace per mile.

Now, if the above records be true in every particular; if Childers and Matchem really performed the feats with which they have been credited, it would appear to be a self-evident fact that the modern racehorse is not by a long way the stayer his ancestor of 140 years ago was. Matchem, it will be seen, on reference to the table, ran more than four miles at a rate of speed which has not been equalled by our fastest Derby winners over a mile-and-a-half course. If Childers and Matchem could run for upwards of four miles at the above average of progression, they must have been not only stayers indeed, but speedy as well, and the decadence of the English racehorse would appear to be a matter not allowing of two opinions. The late Lord Redesdale, too, was of the number of those who held to the idea that the racehorse of about 1866 was not so stout as the horse of a hundred years earlier; but remember what Admiral Rous once said in an article he contributed to *Baily's Magazine*—" My belief is that the present English racehorse is as much superior to the race horse of 1750 as he exceeded the first cross from Arabs and Barbs with English mares; and, again, as they surpassed the old racing hack of 1650. The form of Flying Childers might now win a £30 plate, winner to be sold for £40. Highflyer and Eclipse might pull through in a £50 plate, winner to be sold for £200. This may be a strong opinion; it is founded on the fact that whereas 150 years ago the Eastern horses and their first cross were the best and fastest in England, at this day a second-class racehorse can give five stone to the best Arabian or Barb and beat him from one to twenty miles. I presume, therefore, that the superiority of the English horse has improved in that ratio above the original stock."

It was only in 1885 that the Duke of Portland's four-year-old Iambic, always described as the very worst horse in training, gained an absurdly easy victory over Admiral Tryon's four year-old at a distance of 3 miles at Newmarket. Asil received over four stone and a half. We do not, however, mean to assert that the match in question proved conclusively the true difference in speed between the two breeds; for Asil may not have been as fit as was his opponent; and it may be—we do not say such is the case—that Arabs may require to be trained in a manner differing from that employed for conditioning our own horses. But the match may be taken as showing that the Arab is no match for even a bad specimen of an English racehorse; and the Markham Arabian appears to have been just as great a failure in the time of James I.

Then, again, it is a common cry that we have no stayers in these days. On this point Admiral Rous wrote as follows in the article from which we have already quoted:—" A very ridiculous notion exists that, because our ancestors were fond of matching their horses four, six and eight miles, and their great prizes were never less than four miles for aged horses, the English racehorse of 1700 had more powers of endurance, and were better adapted to run long distances under heavy weights, than the horses of the present day; and there is another popular notion that our horses cannot now stay four miles."

In the time of our ancestors, as the Admiral pointed out, there were but few races over a short distance of ground; now, save at Ascot and Goodwood, we have few long ones; and with so many valuable prizes to be won over courses varying from five furlongs to a mile or a mile and a quarter, it is scarcely worth the while of an owner to forego these stakes, and give his horse a thorough preparation for, say, the Alexandra Plate at Ascot. A horse, like a man, loses his speed if he does the long slow work necessary in training for long distances. In short, the question of staying or non-staying seems to be very much a matter of supply and demand. If the five and ten-thousand pound races were over four miles of ground, the chances are that several horses would be trained thoroughly, with a view to winning those events; but at present it scarcely seems quite fair to give a horse but two or three opportunities of running in a three-mile race, and then to say that he cannot stay, especially as horses stay three and four miles, in steeplechases.

However, we will not venture to dogmatise on a subject which, if it be not quite impossible to prove, is at least very difficult of demonstration. Those who are convinced that the English racehorse has deteriorated, suggest from time to time a re-introduction of Arab blood, which they assert did such good service aforetime. Here it may be pointed out that it was by no means Arab blood exclusively to which we are indebted to the improvement in our blood-stock. Eastern blood of all kinds was imported, as we have already mentioned; and the horse who has been as valuable as any other—the Goldolphin—is said to have been a Barb and not an Arab. As might have been expected, the proposal to again have recourse to Arab blood meets with a good deal of opposition whenever it happens to be brought forward; but at present, and so far as those who breed to race and who breed for sale are concerned, another cross with the Arab does not seem likely to be tried just yet.

It need scarcely be pointed out that it is not to racing men only that our thoroughbred-stock is important. To at least some of those who do own racehorses the racehorse is nothing more than a machine to make or lose money, as the case may be; but to other owners, and to those who have nothing to do with the Turf and its associations, the welfare of the thoroughbred horse means a great deal. He is the foundation of our hunting stock; his blood is represented in our hacks, polo ponies, van horses, and nearly all our harness horses, not to mention the remounts for our cavalry. If it were proposed in Parliament to-morrow to make racing illegal, a thousand tongues would make answer that one of the objects of racing is to improve the breed of horses. This is perfectly true; and by having a good stamp of blood horse we improve, in nearly every particular, hunter, hack and harness horse, in their several types, whatever may be the case on the Turf.

If, then, it be seriously contended that one of the aims and objects of racing is to improve the breed of horses generally; to enable us to boast that we possess the finest horses in the world; to enable us to possess ourselves of the gold of the foreigner; to put money in the pocket of the long-suffering and much suffering agriculturist—if these arguments are seriously advanced, surely the first step towards accomplishing all this is to try and breed a *sound* horse. Yet infatuated people were found to try and bring back Ormonde, regardless of the fact that nearly every veterinary surgeon of note has declared roaring to be one of the hereditary diseases. Surely there is something anomalous in disqualifying for roaring a horse whose chief mission it will be to serve mares at two guineas apiece, while others are willing to pay three hundred guineas for the service of a confirmed roarer like Ormonde, not to mention trying to buy him at a price that would purchase several high-class sound horses. If people like to breed from a roarer, let them; only let them at the same time cease talking about racing tending towards the improvement in the breed of horses.

The importance of breeding from sound parents only has been considerably emphasised since the breeding of hunters and other half-bred stock has been made, though to a comparatively small extent only, a State question. As every one

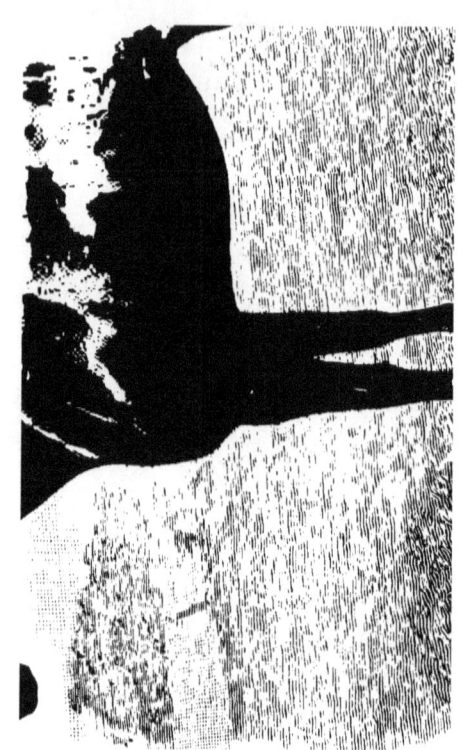

is by this time aware, the money formerly given for Queen's plates is now diverted into another channel, and under the auspices of the Royal Commission on Horse-breeding, an annual sum of £5,000 is now given to subsidising thoroughbred sires which shall serve the mares of tenant farmers at the nominal sum of forty shillings. Now, the most that this arrangement can effect for some time is to place a sound sire within the reach of the average farmer, without giving him the trouble of asking a single question about its soundness. That is guaranteed at any rate as certainly as ever conflicting opinions allow by the fact that the horse is the winner of a Queen's Premium. Yet ever since the Queen's Premiums have been awarded, not a little grumbling on the part of exhibitors has taken place concerning the strictness of the veterinary examination. A ringbone, even at mature age, has caused more than one horse to be discarded; while no matter how perfect shoulders and loins may be, no matter how much bone may be found below the knee, or how flat and clean the legs, any unsoundness in the respiratory organs leads to disqualification.

We are thus confronted with a curious two-fold anomaly. In the first place we have two standards of soundness—one for the sires of racehorses, another for the sires of hunters, hacks, cavalry remounts, &c. Nor does the anomaly stop here. People unacquainted with the Turf might suppose that the greater strictness were exercised over those sires which are destined to beget racehorses; but no! a certain number of breeders are content if they can get a galloping machine, hoping (if we assume that they are not agnostics in the matter of hereditary unsoundness) their horse may be that one which shall not inherit the weak points of his sire or dam. When we come to hunter sires, however (recruited, it must be remembered, from the ranks of blood horses), not to mention Shires, Clydesdales, Hackneys, Suffolks, Clevelands, Yorkshire-bred horses, cobs and ponies, we find soundness in wind a *sine quâ non*. In short, therefore, the readiness with which some racing men will breed from roarers, and the determination of the hunter division to have none of them, seems to be very much as though a man should willingly defile a reservoir, while another person should be very particular as to the purity of the water he draws from the pipe fed by the impure reservoir. Had the Duke of Westminster gone into his picture gallery and sold his ancestors, he could not have been more soundly abused than he was for selling Ormonde; yet that that once great horse left his country for his country's good is as certain as that the Duke would not have parted with him for any money had he, in his opinion, been fit to breed from. It is proverbial that a man, when about to choose a horse, must use his own discretion, but when for breeding purposes he selects an unsound one, he is doing his share towards propagating a race of unsound horses, and, as we said just now, this disqualifies him from pleading that the improvement of the breed of horses underlies the claim of racing to support.

CHAPTER II.

THE HACKNEY HORSE.

THE position occupied by the Hackney horse at the present time, when it is compared with that in which he was placed less than a generation ago, would be absolutely astonishing, even to his most ardent admirers, were they not well aware of the intrinsic worth of the animal upon whose production so many of them are expending not only their time, but vast sums of money. It is, in fact, only of late years that the Hackney has been, if not entirely resuscitated, at all events rescued from the slough of neglect into which the apathy or the ignorance of English breeders had plunged him. Either of these expressions is a hard term to apply to a brother lover of the horse, but yet no more polite one can reasonably be bestowed upon a body of presumably business-like men who had for years ignored the merits of one of the most useful varieties of native horse. That the Hackney has not been popular as an instrument for gambling purposes is certainly very creditable to the breed; but, at the same time, it is quite within the bounds of possibility that, had the English been, as the Americans are, addicted to trotting as a pastime, a great deal of the attention that has been devoted to the thoroughbred would have been lavished on the Hackney.

It is chiefly due to the exertions of the Hackney Horse Society that the horse, to whose interests it is pledged, has emerged from the obscurity which surrounded him but a few years ago, and has once more taken his proper place high up in the list of recognised and popular English breeds. That the Society's efforts have been fully appreciated by the nation is discernible by the fact that Her Majesty the Queen two years ago became its patron; whilst amongst the names of past presidents who have laboured on its behalf, that of H.R.H. the Prince of Wales is prominent in a list which is full of the names of the leading horse-breeders and enthusiasts of the day. Still, royal and aristocratic patronage, inestimable as is its value when bestowed upon a deserving cause, could never have the power to impress the horse-loving public—that extraordinary combination of sentiment and common sense—with any idea of the merits possessed by a horse which, if put to the test, would fail to justify the eulogies bestowed upon him; and consequently the Hackney has been compelled to stand upon his merits. That the horse has amply repaid his friends for their support, the state of the market offers proofs pregnant with silent tributes to his value; and, moreover, this never-failing testimony to an animal's worth—£ s. d.—is supported by the ever-increasing entries at the Society's spring shows in London. Regarding the latter for a moment, it may be pointed out that at the first of these exhibitions, in 1885, there were but 123 Hackneys in the catalogue, whilst at the last, the eighth of the series, held in the spring of 1893, no fewer than 383 entries were secured; and this, let it be added, in the face of a never-ceasing drain upon the resources of exhibiters and breeders by buyers from America, the Colonies, and every country on the Continent. The greatest possible satisfaction must likewise be experienced by every lover of the Hackney, from the conviction that the soundness of the breed is greatly improving as the merits of the horse are more widely recognised and proportionately valued.

Having proceeded so far in the consideration of the position now occupied by the Hackney, one almost begins to fear that readers who have not paid much attention to the antecedents of the breed may commence to labour under the erroneous impression that the subject of this chapter is a horse without a history; but nothing, as a matter of fact, can possibly be further from the fact. The early periods of its existence will shortly be alluded to below; but it is first necessary to refer to the services that have been rendered to the breed by those enthusiasts in East Anglia and in the wolds of Yorkshire, who, for generations past, through good report and evil, have treasured up the old blood that has been left them by their fathers; and whose loyalty and devotion to the Hackney are now bringing forth golden fruit as the fit reward of their staunchness and devotion in the days when the Hackney horse was at a discount. These men stuck to the breed, and bred it pure, not only from feelings of affection, but from that implicit consciousness of its merits which a long association with it had impressed upon their minds; and great indeed must be their exultation in the hour of its, and their, triumph over prejudice and ignorance.

That the Hackney is unfortunately still the victim of both these enemies is an admitted fact that no attempt to explain away can possibly accomplish, but that the horse has now attained an unassailable position, and will live down such attacks as may be made upon him is happily an equally accepted certainty. It is, however, as remarkable as it is regrettable, that the chief imputations upon the Hackney come from professed supporters of the thoroughbred, very many of whom would be highly surprised to learn that the society's Stud Book contains references to, and the pedigrees of, horses that were foaled as far back as the middle of the last century. Consequently, it is surprising that such sticklers for blood should express disdain at animals, in whose veins the blood of the Darley Arabian and his illustrious successors most undoubtedly flows, but it is perhaps within the limits of possibility that the attention that such breeders of hunters as Earl Spencer and others are bestowing upon the Hackney, has excited a feeling of rivalry within their breasts. Admitting for the sake of argument—and of argument only—that the thoroughbred is all that is pure, and that the Hackney is but half-bred, it must be confessed even by the detractors of the latter that he is a mongrel with an exceptionally long pedigree, in many cases as far reaching as that of the first mentioned horse. It can scarcely be maintained, moreover, that such sires as the Darley Arabian or Godolphin Arabian were in the zenith of their fame only put to galloping mares, in fact, abundant proofs are forthcoming to the contrary; neither can it be contended or substantiated by evidence that other light mares, besides gallopers, were not highly prized by horse breeders in the eighteenth century.

Mr. Henry F. Euren, the energetic secretary of the Hackney Horse Society, and an enthusiast upon all questions connected with pedigrees, has satisfied himself by a reference to the files of the *Norwich Mercury*, of the breeding of the original old Shales, a horse which is regarded by modern hackney breeders as the foundation stone of the Stud Book. Shales, according to advertisements which appeared in the *Norwich Mercury*, in April, 1772, and March, 1773, was the sire of Scots Shales, who was serving at a fee of one guinea a mare, and a shilling to the groom, and is stated to be " by *a son of Blaze;* Blaze, by Childers out of a well bred hunter mare." Blaze was foaled in 1733, and was by Flying Childers, dam by Grey Grantham by Brownlow Turk out of a mare by the Duke of Rutland's Black Barb. Of the many sons of old Shales, two at least, Driver and Scots Shales, in turn became pillars of the Stud Book, and to the former of these horses many—very many—of the best Hackneys of the present day owe their origin. For instance, Mr. Philip Triffit's great sire Fireaway was by Achilles (Hairsine's) by Fireaway (Scott's) who was got by Fireaway (Ramsdale's) by Fireaway (Burgess') by Fireaway (West's) by Fireaway (Jenkinson's) a son of Driver, from Mr. T. Jenkinson's mare by Joseph Andrews by Roundhead. Other instances—almost innumerable—are forthcoming to prove that Hackney breeders of the past,

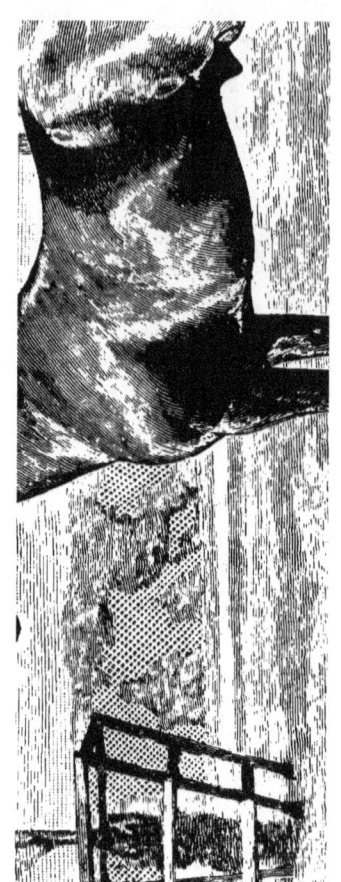

although unfortunately denied the benefit of a Stud Book, were not regardless of the value of a pedigree, and it would be as insulting to the memories of these departed breeders, as it is opposed to the dictates of common sense to maintain that they were neglectful of the breeding of the mares from which they raised their stallions. That a certain amount of obscurity must always exist concerning the authenticity of ancient pedigrees is a melancholy fact which all experienced breeders are compelled to recognise, but it can scarcely be contended with any reasonable show of justice, that the pages of a printed Stud Book must invariably be correct, if all verbal or traditional testimony, such as handing down the pedigrees from father to son, is necessarily wrong. All breeders must admit that changelings have existed amongst thoroughbreds, and consequently, this being so, the records of their Stud Book might as well be described as being unreliable, and with as much justice as Hackney pedigrees are assailed by those who cavil at the fact that every detail of early breeding is not set down in black and white.

Reverting, however, to the antiquity of the Hackney as a recognised breed, it may be stated that most ample testimony is forthcoming in support of this assumption, for which the writings of many recognised authorities are witnesses. John Lawrence, whose exceedingly practical contributions to the literature of the eighteenth century were compiled under difficulties which will be fully appreciated by modern authors, makes frequent allusions to a breed that was in all its principal characteristics identical with the Hackney of the present day. From a perusal of the "Philosophical and Practical Treatise on Horses," published by him, we find that "in former days" the horses for the saddle were nags, amblers, pacers, stirrers, trotting horses, hobbies, great horses, or horses for the buff saddle, hunting horses, coursers, racehorses; whilst he proceeds to add, the appellatives in "present use amongst us," are road horses, riding horses, saddle horses, nags, Chapman's horses, hacks, *hackneys*, ladies' horses or pads, hunters, running horses, racers, race horses, gallopers, welter horses, managed horses, chargers, troop horses, post hacks or post horses, trotters, cantering hacks or canterers, horses which carry double, cobs, galloways, ponies and mountain-merlins. There is fortunately no necessity for analysing the above rather intricate list of varieties and sub-varieties in the present instance; the object held in view in making the quotation being fully served by the fact that the existence of the Hackney as a recognised breed a hundred years ago is amply demonstrated. John Lawrence, moreover, in his "History of the Riding Horse" again alludes to the *hackney* as a recognised and valued variety, and connects him with the roadster—a designation which apparently is there used for the first time in English equine history, although so far back as the year 1600, Hakluyt makes use of the expression "roader" in his collection of travels, which goes a long way towards proving that the title was in vogue in America at an even earlier date. There can, however, be no reason for doubting that the designations, hackney and nag, were interchangeable terms when applied to horses during the past century, and certainly they have remained so ever since, as in many parts of the country they are applied indiscriminately to animals of the same variety. According to Mr. H. F. Euren's carefully compiled and most valuable introduction to the first volume of the Hackney Horse Society's Stud Book, the expression nag is the oldest surviving appellation for the active riding horse in this country, and the word, he informs his readers with evident correctness, is derived from the Anglo-Saxon word hnegan, to neigh. Subsequently, Mr. Euren proceeds to state, the Normans when they took possession of this country, introduced their own word, haquenée or hacquenee, which was recognised in England so far back as the year 1303. As a proof of this, Mr. Euren quotes the following extract from the "Vision of Piers Plowman," written in 1350, "ac Hakeneyes hadde thei none, bote Hakeneyes to hyre." The adoption of the expression

Hakeneye unquestionably proves the antiquity of the word, and the general use made of it in England at that date, but unfortunately there are no details forthcoming to inform modern seekers after light what sort of animal it was that was referred to by it.

That East Anglia was, if not actually the home of the Hackney, at all events the locality in which horses of a similar character were very highly esteemed so far back as the fifteenth century, is made obvious by the allusion made by Dame Margaret Paston, who in writing to her husband informs him that "there be bought for you three horses at St. Faith's Fair, and all be trotters, right fair horses—God save them—and they be well keeped." From the above quotation Mr. Euren very reasonably argues that the absence of any further comment upon the appearance of the trotters proves that at that time there was a recognised type of horse, in Norfolk at all events, which was commonly known as a "trotter."

In those old times, as now, a good deal of difference appears to have existed as regards the value of horses, for one learns that in 1462 Lord Howard paid £1 16s. 8d. for a grey nag to send to the French King; whilst eight years later it is recorded that upwards of £13—about £75 of our money—was the price paid for a good animal. Of Sir John Falstolfe it is also recorded that in 1435 he bought two weight-carrying saddle-horses at Yarmouth at £11 2s. 6d. apiece, which may be regarded as a good price for the age in which the transaction occurred.

The reign of that excellent monarch and good sportswoman, Queen Elizabeth, was undoubtedly a glorious period for authors of all descriptions. Everyone, of course, knows that the immortal Shakespeare flourished in those days; but few possibly are aware that the first book in English upon dogs was written by her physician, the famous Dr. Caius, the founder of Caius College, Cambridge; and also that the first English book on horses by Master Blundeville, of Newton

Flotsham, in Norfolk, was published during the time she occupied the throne. The last-named writer was evidently an enthusiast, and the completeness of his work is surprising, when the difficulties under which he laboured come to be considered. He was a great advocate of scientific horse breeding, and manfully maintained the rights of Englishmen to exercise their own discretion regarding what variety of animal they should keep. For breeding serviceable horses Blundeville recommended mares that were "of an highe stature, strongly made, large and faire, and have a trotting pase," for as he holds, "it is not meet for divers respects that horses for service should amble."

The size of the old-fashioned horse is a subject upon which very little light has been thrown by writers, but the Duke of Newcastle, who published a work upon equine subjects in the reign of Charles II., expresses the opinion that "there is no fear of having too small horses in England, since the coolness and moisture of the climate and the fatness of the land rather produce horses too large." Such assertions are, however, of far too vague a nature to afford any reliable data for estimating the stature of horses during the reign of the Merry Monarch, but early in the seventies there is, fortunately, more reliable information forthcoming to show the size of the stallions which produced the mares which a few years later on formed the foundation stock from which the modern Hackney was developed by the infusion of Arab blood. We learn from an advertisement which appeared in the *Norwich Gazette* in 1725, that a grey stallion, standing 14 hands, was at the service of the public, whilst two years later an announcement appears in the *Norwich Mercury*, giving the description of a large stout coach *gelding* of 15 hands, which shows that the stallions, though they appear to have averaged from 13 to 14 hands, were still capable of getting something taller than themselves. Later on, in 1729, one finds that a stallion—"an Arabian," but more probably a half-bred horse—was in Norfolk, and as the century advanced in age so the height of the horses

apparently increased. Attention may, however, be here directed to the fact that up to a comparatively very recent date the Hackney classes at the annual shows of the Royal Agricultural Society of England were only open to horses *not exceeding* 15.2 hands in height, but happily this absurd restriction has been removed. Early in the present century there must have been some very big Hackneys about, as a reference to the Society's Stud Book shows that Fireaway (West's)—known as Silver-tailed Fireaway—who was foaled in 1807, stood 16 hands; and he in turn begot the dappled grey Phenomenon Fireaway, whose height was 16.2 hands, out of a mare by Hazard. West's Fireaway, it may be mentioned, was a g.g grandson of Shales the Original, and being a Norfolk bred horse, as may be supposed, came out of a trotting mare by Pagan by Spectator, Pagan's dam by Blank by Godolphin Arabian, Spectator by Arab by Alcock's Arabian.

It is somewhat remarkable, however, to note the circumstance that most of the successful stallions, both as sires and exhibition horses, of the present and past generations, have stood 15.2 hands. Fireaway (Triffit's) was of this height, as was Mr. George Bourdass's Denmark, and the latter's scarcely less illustrious son, Danegelt, is also 15.2. Mr. W. Flanders' celebrated horse Reality, the absolute winner of the first of the Elsenham Challenge Cups for stallions which were presented by Sir Walter Gilbey to the Hackney Horse Society, is likewise of this height, as is Star of the East; whilst the famous black Prickwillow (Tice's), and his equally renowned son Confidence (D'Oyly's) were also 15.2 at shoulder.

Having thus endeavoured to briefly trace the existence of the Hackney horse, so far as his origin and stature are concerned, to the remotest period in the history of the breed, the attention of the readers of this chapter may be directed to the purposes for which the horse was primarily produced, as thereby some light may perchance be thrown upon what may be regarded as the correct type of Hackney to select. At this point, again, Mr. Henry F. Euren, in his invaluable production, the introduction to the first volume of the Stud Book, will be found to have expressed a closely reasoned opinion on the subject, the truth of which, especially as he quotes from co-existent authorities, it is impossible to question. Mr. Euren states roundly that "the construction of railways had a speedy effect on the breeding of Hackney horses," and this circumstance doubtless affords some explanation for the lack of popular support the breed received a couple of generations ago. Before the benefits of Stephenson's great invention were offered to the public, farmers in all parts of the country were compelled to go to market on horseback, and, in the words of Mr. Euren, it was "no unusual thing" for them to ride fifty or sixty miles a-day to and fro. Now the British agriculturist, whatever his other imperfections may be, has always been regarded as a solid man—at all events, from an avoirdupois point of view; and consequently it must have been a powerfully-built horse that was used to carry him. The animal, moreover, could not have been a sluggard, for time is valuable when marketing transactions are on the *tapis;* and, therefore, it must generally be conceded that the horse the farmer bestrode must have been fast as well as strong. In making use of the expression fast, it is perhaps desirable to state that the trot is not the only action in which the Hackney should excel, for the style in which he walks is a point that should always be taken into consideration by judges of the breed. Few horses are capable of negotiating a long journey at a trot with a heavy man upon their backs; and even if they were, it is to the last degree improbable that many men would be forthcoming who were able or willing to ride for twenty miles at this pace. It is, therefore, necessary when selecting a Hackney to bear this in mind, as none of the old-fashioned farmers would have cared to lose time when easing their horses, owing to the animal they rode being a slow walker.

We are, therefore, pretty well able to estimate what the

action of the Hackney was like during the earlier years of the present century, whilst, as regards a still more antecedent period, there is plenty of evidence to hand to bear witness to the abilities of the then existent trotting horse. For instance, it is recorded of old Shales' son Driver that he was the sire of a mare that trotted 15 miles within the hour, carrying 15 stone, whilst the old horse himself is credited with having accomplished 17 miles within the same period of time. According to an advertisement in the *Norwich Mercury* in February, 1806, Pretender, who was therein alluded to as being for sale by public auction, is stated to have done his 16 miles within the hour, with 16 stone upon his back, when five years old; whilst Read's Fireaway won the second prize given by the Agricultural Society to the best trotter, in the year 1801, and after the victory trotted one mile in 2 mins. 49 secs. At a much earlier date, however, there were records forthcoming to prove the ability of the English trotter, amongst which is one held by Phenomena, a mare by the trotting stallion Othello, out of a Norfolk trotting mare, who is credited with a record of 17 miles in one hour. Phenomena, although under 14·2 hands, when 12 years old, in the year 1800, trotted 17 miles in 56 minutes in the Huntingdon Road; and, the performance being questioned, repeated the achievement—in fact, excelled it—by negotiating the distance in a few seconds under 53 minutes. Subsequently she was matched to complete $19\frac{1}{2}$ miles within the hour for a stake of 2,000 guineas; but her opponents paid forfeit when they discovered that in a trial she had trotted 4 miles under 11 minutes. It is also recorded of this famous Hackney that, when 23 years old, she did her 9 miles in $28\frac{1}{2}$ minutes. Phenomena owned several masters in the course of her career, amongst them being the Duke of Leeds, who was content to pay 1,800 guineas for her when in the zenith of her fame; but her old age was not respected by those she served, for she was eventually disposed of in one of the London sale yards for the wretched price of seven pounds:

but she lived to serve both Mr. Daniel and the Rev. Dr. Astly well, eventually dying suddenly in the last-named gentleman's possession. A grandson of Phenomena appears in the Hackney Society's Stud Book under the name of Phenomenon (Jacobs), 578.

So much for a brief review of the past history, pedigree and performances of the Hackney horse, which review might readily have been greatly extended had the amount of space available for the present chapter been larger than it is. It is, however, now necessary to offer some observations upon the modern horse, not with the idea of comparing him with the heroes of bygone generations, but rather with a view to describe the position which he occupies, the uses to which he may be put, and the form in which he most frequently occurs. That many Hackneys of the present day have lost a great deal of the old character is a fact that must have impressed itself somewhat strongly upon the minds of those who have studied the ancient and modern types of animal, but it must, at the same time, be equally apparent to all that such alteration as has been effected is easily to be accounted for by the fact that the present horse is put to far different duties from those upon which his ancestors were employed. No heavily-built farmer—or, indeed, any other person in full possession of his senses—would, in these enlightened days, ever think of escorting his buxom better-half to market perched on a pillion behind his saddle; and even if such an idea was to enter the head of some eccentric individual, it is to the highest degree improbable that the lady would be a consenting party to the arrangement. Consequently, there can be no serious objection raised against the contention of Mr. H. F. Euren that the advent of railways affected the production of the Hackney horse. Long journeys are now every whit as great events in a man's career, if conducted upon horseback, as they were half-a-century ago, when negotiated by train; and therefore it is not surprising that the more powerfully-built stallions have of late years been less favourably regarded by breeders than their more bloodlike relatives. At the same time, there can be no denying the fact that when quality, and quality alone, is looked for in the Hackney, a great deal of the horse's real value is being lost sight of in the search. A coarse, underbred looking animal should always be avoided by a Hackney breeder, unless so be that the stallion's pedigree is unimpeachable, and his services are required for a light or weedy mare; but even then there is the risk in introducing the blood of a plain-looking sire into a strain, as, however good he may be, a coachy-looking youngster will always require a great deal of selling as a saddle horse. Consequently, in the desire of avoiding the Scylla of too much quality, which so frequently entails weediness, a breeder must use caution not to wreck his enterprise upon the Charybdis of "timber," which as often is attended by so large an amount of substance as to bring coarseness in its train. It will, therefore, be observed that the class of Hackney that not only represents the old-fashioned type the most accurately, and which, moreover, is certain to command the highest price, is a powerfully-built, short-legged, big, broad horse, with an intelligent head, neat neck, strong level back, powerful loins, and as perfect shoulders as can be produced. Such details as feet and muscle need scarcely be discussed at present, for it is obvious that any animal that is deficient in the latter would be unable to do his work; whilst, however good a horse may be in other respects, if his feet are malformed, or too small to carry him safely, he must clearly be worthless.

Commencing with the head, it may be said of this most important Hackney point that it should be, comparatively speaking, wide at the jowl and taper gently towards the muzzle, the eyes being of a good size, so as to, in conjunction with the shape and dimensions of the head, convey into the mind of the observer, an impression of strength, intelligence, and courage, combined with quality. A very small effeminate-looking headpiece is almost as objectionable in a Hackney as a heavy one, the former usually denoting an absence of resolution combined with softness; whilst the latter is frequently accompanied by a sullen, dogged temperament, which is altogether opposed to the true nature of the horse. In short, in dealing with the head of this class of animal, the general symmetry of its conformation should be studiously regarded, and, therefore, although a small head is always an attraction, and most properly so, it is ridiculous to encourage one that would suit a 14 hands horse when it appears in a Hackney six or eight inches taller. The ears should be small and pointed, although animals have won prizes adorned with organs of hearing that would scarcely have disgraced a lop-eared rabbit, whilst a neat though well-defined crest is undoubtedly a great beauty in the case of a stallion.

The neck should be of fair length, nicely bent, and rather thick at the setting-on, though free from coarseness; whilst the chest must be wide, and let down behind the forearms so as to allow plenty of space for the heart and lungs.

The shoulders of the Hackney, as in the case of all riding horses, should be free from all that superfluous lumber which only brings coarseness in its train; they should lay well back, and the bones should be long enough, forming as they do the attachment of the muscle *serratus magnus*, which connects the fore limbs and trunk. If these bones are small the muscles must necessarily be short, and long muscles alone can afford that flexibility and liberty of shoulder action which is so characteristic of the Hackney.

The back must be long enough to allow plenty of room before and behind the saddle, and also very level and strong, whilst the loins should be compact and the quarters long and as powerful as possible without being coarse or of that coachy type which is so distasteful to many judges. The middle piece of the Hackney is very level above and below, the ribs being well sprung and the back ones of a nice length, so as to provide those indications of strength which are always to be eagerly sought after.

The fore legs should be short and very powerful, the arms being big and muscular, the joints large and the bone below the knee plentiful and flat; whilst the pasterns should be of a fair length, so as to yield elasticity in action, and the feet of good size and placed straight on under the legs.

The hind legs should possess powerful sound thighs, strong, well-bent hocks, and ample bone, whilst the tail, which is set on rather high, should be carried gaily when the horse is fully extended.

The above short delineation of the leading points of a Hackney may be taken as detailing the chief characteristics of a successful show animal, but the merit of the description lies, in the words of Captain Cuttle, in the application of the same. So many different people view the same features from different points of view, that in arriving at a decision upon the properties of a horse, two judges, although entirely in unison as regards what they want to find, will often entirely disagree as to what points importance should be attached to when called upon to give expression to their ideas in public. Two excellent authorities have even been known to differ upon the comparatively easy question of bone, the one avowing that although the horse was just a trifle light below the knee, he still possessed enough for all practical purposes, whilst his colleague declared that the animal was a weed and wholly unworthy of a prize. Some explanation is, therefore, forthcoming by the publication of this incident, that will readily account for the reversal of the decision made by one judge, by another gentleman whose views are known to be in harmony with his views. Briefly, therefore, it may be taken that in judging Hackneys, two experts, although possessing identical opinions regarding the points of the breed, may yet fail signally in attempting to reconcile their practical application of the ideas they hold.

The structural development of the Hackney having been considered, a no less important property belonging to the breed must now be referred to, as a trotter, however speedy he may be, is certain to receive but very scant attention from judges in the show-ring if his action is defective. Any horse, in fact, loses more than half his value if he fails to "move," and there is just as great a difference in the action of animals as there is variety in the gait of men. The Hackney in one respect stands alone amongst light horses, as he is, so to speak, one movement short, for he rarely gallops and relies entirely upon the trot and walk as methods of locomotion. The former of these gaits is so entirely characteristic of the breed, that one can readily believe the truth of the statement that many a Hackney can trot faster than he can gallop; but be this as it may, it must clearly be understood that no pacing —by which expression the moving of both legs on the same side of the body simultaneously is implied—or ambling is permissible in the case of a Hackney, whose trot is a trot pure and simple, and unassociated with any eccentric exaggeration whatsoever. In fact, the one, two, three, four, of a true-actioned Hackney as he pounds along is veritable music in the ears of an enthusiast, and no breeder of the horse will ever, it is certain, be prevailed upon to try to do without it.

Hackney action, however, like everything else in this world, varies in quality, and in this respect, as in all others, judges have been known to differ very materially in their views, though none of any position have ever been known to give a prize to a palpably shoulder-tied horse, and few possibly would ever dream of breeding from the same. As in the case of every other breed, the shoulders, knees, and pasterns are all called upon to contribute to the front action, but in no variety do the shoulders exercise such important functions as in the Hackney, and hence the importance that has been bestowed in the description of his points given above to the length of their bone, which ensures a good and elastic *serratus magnus* muscle. If a horse does not possess this he can never be free-shouldered, the result being that when he bends his knees, as he should do, it becomes a case of all action and no go, which is absolutely useless for every practical purpose, for if a trotter does not get away in front all the beauty of his going is completely lost. Consequently plenty of freedom and liberty about his shoulders is to be regarded as being a *sine qua non* in the selection of a Hackney, not only on the grounds of the additional grace which such a conformation imparts to his movements, but because, assuming that his pasterns are also good, he is far less likely to knock his feet to pieces against the ground when fairly let go, than he would be if he only lifts his knees up and smashes them down again in an almost perpendicular position.

A good knee action is of course essential to the success of any horse, and the higher they are raised—assuming always that they are straightened again in time to effect that most peculiar poise which many of the best Hackneys show just before their fore-feet reach the ground—the more generally the action of the horse will be admired. Exaggerated knee action is usually only obtained at the expense of freedom at the shoulder, and few things are more irritating to witness at a show than a good-looking horse lifting his knees nearly up to his muzzle, and then putting his feet down in almost the identical place from which he raised them.

Regarding the use that a horse makes of his pasterns when he trots, it may be explained that a short upright joint promotes concussion, and naturally not being so flexible as a long springy one, is not so readily bent back and then straightened when the horse is trotting. The feet in the case of the animal which has good pasterns are in the majority of cases picked up and put down smoothly and levelly, without any of that dishing or throwing from side to side which simply spoils the action of an otherwise fine mover. The pastern joints are, therefore, it will be seen, valuable co-operators with the shoulders in providing the Hackney with the desired front action, and when properly moved by the animal will be found to assist in bringing the feet down with that comparative gentleness which contributes so largely to their remaining in good condition.

The stifles and hocks, and of course the pasterns to a smaller extent, are the joints which regulate the back action of a horse, the hocks bearing by far the most important share of the movement when the animal is on the trot. Naturally the stifles must be moderately bent, else no pace will be secured, but the chief merit in a Hackney's hind action is the style in which he moves his hocks. When these are first well bent, and then brought nicely under his body, a great amount of extra ground will be covered, and if so be that his quarters are free from all superfluous lumber, and his gaskins powerful, his propelling power will then be tremendous. No horse, Hackney or otherwise, can ever get over the ground at a reasonable pace if he leaves his back legs behind him, added to which it deprives the action of a Hackney of the regularity which is its greatest charm, if all the four limbs do not move in unison. Many good Hackneys possess a tendency to go very wide behind between the hocks, which is unsightly, although the act is frequently accompanied by undoubted speed. It has been stated that a barrow might almost have been wheeled between the hocks of the Flying Dutchman when he was fully extended—but then it must be remembered that the great horse in question was not a trotter, neither does the maker of the observation appear to have remarked that his action was improved in appearance by the habit. The development of a propensity to go too wide behind, although an eyesore, need not necessarily prove that a Hackney is unsound, and moreover, it is probably in many instances the result of a mistaken principle in schooling a young horse, by which he is encouraged to over exert himself before he is old enough to take any liberties with his action. Horses that turn in their hind feet ought never to be passed, whilst in cases when the latter are turned out, the animal will be found to be more or less cow-hocked, which likewise is a very serious fault in an animal who should stand square and move truly above all things.

The principal gait of a Hackney—the trot—having been

discussed, a few observations may now be directed to the scarcely less important walk, which is so priceless a blessing when added to the other virtues of a good saddle horse. There is no compromise about the true Hackney walk, as the possessor of it steps out all round, throwing his front legs well before him, and his back ones right under his body in a style that makes his stride enormous. Many of the leading Hackney stallions of the present day are by no means good walkers, and more's the pity, as if they possessed the gift and transmitted it to their stock it would add most materially to their worth as sires, and to the value of the youngsters in the market. Without the least desire to draw invidious distinctions between the respective merits of different horses, it is impossible, when alluding to the walk of a Hackney, to avoid drawing the attention of our readers to the grand walking action of Mr. Tom Mitchell's chestnut Ganymede, who was bred by Mr. John Wreghitt, in 1888, by Danegelt, from a mare by one of the numerous Phenomenons whose names adorn the pages of the Stud Book. A superb mover in both paces, Ganymede is simply the perfection of a walker when exhibited in proper condition, which has not always been his fate. So long as breeders pay attention to the walk of their horses, they are pretty certain to find a sale for the animals, as most persons who ride have reason to appreciate the value of an animal who, when trotting is impossible, can get over the ground at a good pace.

It is, of course, a self-evident fact that many Hackney owners never have any occasion or desire to ride their horses, and in selecting an animal for harness purposes, the great difficulty in finding the right sort of shoulder is materially diminished. A shoulder for a riding horse must necessarily be long if the equestrian expects to secure a comfortable mount, but neither the length nor the slope of his shoulder is a matter of such vital importance in the selection of a harness horse, provided always that each of these required points is sufficiently developed to ensure that freedom of the shoulder that is so essential to good action. That the Hackney is destined to become ere long a very popular animal in the capacity of a coach horse, was made evident some time ago, when Lord Hastings disposed of a pair of gigantic bays, which had drawn his carriage upon state occasions, and which were still pure-bred Hackneys. Their great stature, however, as may be seen by a reference to the observations which have already appeared above in reference to the height of Hackneys, should not be a matter of surprise to breeders, who, now that experienced persons on all sides are doing their utmost to improve the horse, may reasonably expect to find an increase of stature amongst the members of their studs. Good food and careful housing must always lead to the development of an animal's frame, so long as such benefits are not carried to extremes, but when a certain height at shoulder—say 15.3 hands—is reached, every additional fraction of an inch will be a difficult matter to secure. Exceptional horses, such as Lord Hastings' bays, will always be appearing, but although the occurrence of such giants will be more frequent, it will be many a long day before 16.1 or 16.2 hands comes to be regarded as anything but an exceptional height in a Hackney.

The question of temper is always an important one in the selection of a horse for business or pleasure purposes, and it will usually be found that the more docile an animal is, the greater will be his courage when the pinch comes, whilst his value will, of course, be far greater than that of an evil-dispositioned beast, who at any moment is liable to injure his owner and stable companions in one of his displays of temper. Fortunately, for the reputation of the breed to which they belong, Hackneys are the most amiable of horses, and the appearance of a vicious stallion in the show ring is almost unknown amongst exhibitors, but, unhappily, it is neither the nature nor the breeding of many a savage horse that has made him what he is. Bad breaking, the tricks of shallow-pated grooms, and the teasing some youngsters receive from the idle loafers—adult and juvenile—who contrive to gain admission

Hackney Mare, 1671 Lady Wilton 2nd.

to the stables, are each and all responsible for many a ruined temper; whilst the vagaries of a wooden-headed owner, who, although possessing the hands of a quarryman, is of the opinion that, because he has paid a long price for a spirited horse, he must necessarily be able to drive or ride the animal, have contributed to the ruin of good colts innumerable. If the owners of a valuable Hackney would only realise the fact that they have no hands, when they are unfortunate enough to be so afflicted, they would never lay themselves out to incur the ridicule that is bestowed upon them by the spectators of their folly, but as matters go they usually are incapable of recognising their own imperfections, and thereby succeed in ruining the prospects of their horse. So perfect, however, is the temper of most Hackneys that many breeders who are also farmers ride their stallions regularly about their fields when the men are at work, and beyond all question, when the rider is a horseman, the animal is benefited by the useful amount of healthy exercise thus afforded him.

The disregard to the condition of their stock, and the circumstances under which the young ones are reared, that is evinced by some owners is really appalling; and creates surprise amongst those who are acquainted with what the animals go through. For instance, a stallion, whose action should be one of his strongest recommendations, is sometimes brought up to the early show, just before the covering season commences, so loaded with fat that he can scarcely move; and yet the unfortunate animal is expected to take a prize, and subsequently to travel the country and foal his mares! He probably fails in both attempts, and thereby loses his reputation for looks and as a sire, when under ordinary circumstances he would have been able to have done all that was desired of him. Then in the case of yearlings, they are blown out upon cake and boiled beans, and other stimulating diets until they give an old-fashioned admirer of Hackneys the impression that they have been prepared for slaughter and not for show; whilst most of all the object of this ridiculous treatment of the youngsters is that they are fed up to look big and take a prize. It would be interesting to know how many of these precious juveniles have developed into good horses by the time when they arrive at a full mouth, but certain it is, that many disappear from the scene, and are heard of no more, whilst others get beaten time after time by opponents who, though behind them as yearlings, had not been forced, and who, therefore, have come on whilst they themselves have deteriorated. Doubtless, some yearlings are sold to go abroad, or into remote parts of the country after they have scored an early success or two, and with their blushing honours thick upon them have succeeded in gaining a reputation in the locality that has served them at the stud in after life, nor must it be assumed that every prize yearling has been the victim of injudicious pampering and a heated stable. Fortunately, all owners are not short-sighted enough to adopt tactics with their youngsters which will jeopardise the success of their future career, and Mr. Henry Moore, of Burn Butts, the owner of one of the most successful Hackney studs in the country during the past eight years, is a notable instance of a breeder who brings his young stock up hardy. The practice of this gentleman is to let his horses lie out all the winter, the result being that they come up to spring shows as rough as bears and as hard as nails.

It is, however, questionable whether this system of leaving horses, and especially young ones out during the existence of cold weather is a desirable one in all cases, as constitutions differ, and it is not every owner of a stud who possesses the knowledge when to bring an animal that is suffering in. The whole question of turning horses out to grass is one that appears to be very imperfectly understood by the average horse owner, who, in many instances, never thinks, or if he does give the subject a thought, is incapable of forming an opinion of what he is about. How often for instance, does one notice during the summer months the horses of some non-reflecting neighbour turned out during the day and brought back to their stables of a night, an arrangement which effectually secures their being tormented by flies and the sun when in their paddock, and deprived of the benefit of all the refreshing morning dews, which cool their feet and render the grass moist and toothsome. Surely, therefore, this system of summering horses requires some revision when any benefit is expected to be derived by the animals from their short emancipation from the drudgery of active service; and it is to be trusted that owners who desire to improve the condition of their animals will devote a portion of their leisure to the consideration of the requirements of the latter.

The above few lines, however, must be regarded as being rather in the nature of a parenthesis, as the question of summering and wintering horses is scarcely one that can properly be dealt with upon its merits in this chapter. It was, nevertheless, introduced as being a matter which has much to do with the constitution of a horse, it being certain that any animal, which is either pampered when young or improperly treated when old, is never likely to do itself justice at the stud, even if its success in the show-yards is not seriously prejudiced. A Hackney, above all horses, should be of a robust and vigorous constitution, for whilst admitting that the exigencies of the age, which regulate the great question of supply and demand, require a somewhat lighter animal than was sought for formerly, it must always be remembered that a Hackney without substance and power has lost two of the great properties possessed by the breed. It is almost to be feared, however, that this demand for quality may influence judges into paying too much attention to animals whose chief merit is their style and blood-like outline, to the detriment of the old-fashioned type which made the reputation of the horse, and will sustain it for all time, if given due encouragement. That some judges go for blood and others for substance, is rendered obvious by the presence in the same prize list of animals of both the heavy and light types. Anomalies of such a description are always to be deprecated as being confusing to the public, and bewildering even to experienced breeders who very naturally may enquire what sort of horse it is that judges want.

In selecting a Hackney mare, the seeker after the right sort of animal should always be on his guard against permitting a very natural regard for that most elastic of all equine virtues—quality—to override his judgment, and cause him to give preference to an animal that looks like being three parts thoroughbred, over a long, low, heavily-boned mare, who knows how to use her shoulders and bend her hocks, and whose pedigree alone should prove that she is bound to throw a Hackney. In the expression of this opinion the writer does not desire it to be imagined for an instant that he is unappreciative of the value of blood and style about a Hackney, but at the same time having often wondered why it is that some judges give prizes to mares that are almost ladies' hacks, he ventures to suggest with all diffidence that the rage for quality may be carried a little too far, as there is always a chance of getting too much even of a good thing.

If a brood mare is too much on the leg and deficient in bone the probabilities are that a plain, heavy horse will be selected for her with the idea of ensuring that plenty of substance shall be about the foal. This is perhaps an inevitable result of breeding from light mares, but a coarse sire is almost certain to transmit some of his plainness to his stock. Thus, as his sons are likely to be bred from in time there is always a probability, unless their mares are all most carefully selected, of many of their foals throwing back to the plain grand-sire, and, consequently, a light mare may prove a medium for introducing coarseness and loss of quality into a strain, simply on account of the efforts that have been made by her owner to counteract her own defects. Still, necessity knows no law, and it often occurs that an owner finds himself in the awkward position of having to breed from an animal which he does not really fancy. In such a case, if he is a wise man, he will be very careful about introducing the blood of the

result of the cross into his own strain, for it is always safer to sell a doubtful horse or mare than to breed from it. The master of a stud is always to be envied, therefore, when he is able to breed from animals where there is type and character on both sides, as even though he may not always be fortunate to discover super-excellence in every foal, he may feel reasonably certain that his younsters will not be bringing in faults that may require years to breed out. No experienced owner will, of course, ever dream of sending a mare of any kind, let alone a valuable one, to a horse that he has not seen, or whose pedigree he has not satisfied himself is right in all respects. All men, however, who raise foals are not to be regarded as breeders in the highest acceptation of the term, and will take a nomination to a stallion simply because he has won a prize, and because they think his stock will sell. Such people are, nevertheless, acting most unwisely even in their own interests, for blood is always thicker than water, and pedigree is sure to tell in the long run. The Hackney would not be the horse he is if the old breeders, whose staunchness in the past has been the means of saving the breed from extinction, had not paid attention to details, and a happy-go-lucky system of stud management will never pay in the long run.

There is no possible explanation forthcoming to account for why one horse or mare should be a success in the stud, and their own brother or sister a perfect failure; and, therefore, the safest way to proceed is to put, so far as possible, one's best mares to tried horses. Still, as the poet Horace, who was a bit of a farmer himself and a horse breeder, doubtless, very truthfully observes, "*Fortes creantur fortibus et bonis*," and the maxim is as applicable to Hackneys as it is to any other animals under the sun. Having no desire to delve once more into the dim traditions of the past, the writer does not propose to go further back than the last generation to prove the accuracy of Horace's observation, and will, therefore, content himself with selecting Mr. George Bourdass' Denmark H.H.S.S.B. 177, and Mr. Philip Triffit's Fireaway H.H.S.S.B. 249, as illustrations to go by. These two grand old horses, whose memories will for ever remain green in the minds of those who love a Hackney, were both big prize winners twenty years ago and more, but their victories in the show ring are as nothing compared with the services they subsequently rendered at the stud to the breed which they adorned whilst alive. The value of a Denmark mare is notorious amongst Hackney men, and the old horse also sired the winners of two of the Hackney Society's Championships—Candidate and Connaught—the former of which got the champion M.P., a dual winner of this honour.

Nor are the big winners of modern times one whit behind these two old champions in stamping their quality upon their get, as witness the vast number of prizes that find their way to the sons and daughters of Mr. W. Flanders' Reality, the absolute winner of the first Elsenham Challenge Cup, Mr. Burdett-Coutts' Candidate, the second winner of that event, and last, but by no means least, of Mr. Henry Moore's defunct chestnut, Rufus, who, like Reality, succeeded in winning the second Elsenham Challenge Cup for his owner. The sudden death of this great horse may be regarded by breeders as being little less than a calamity, as amongst his stock that have appeared there is scarcely one that has failed to be a credit to his illustrious sire— one of the best Hackneys of modern days. All the get of Rufus are long and low, with plenty of substance, and a heap of Hackney character and quality about them. Thanks, therefore, to Mr. Moore's great chestnut, admirers of, and believers in, the old fashioned Hackney should soon be able to recruit their studs by a dash of Rufus blood, the value of which should be inestimable, as he left a good crop of foals behind him, when he was cut off in his prime a short time ago. Rufus, although his show career began and ended whilst he was an inmate of a Yorkshire stable, was by birth a Norfolk horse, having been bred by Messrs. Peacock and Sons, of Brandon, by the chestnut Vigorous, dam Lady Kitty, by Jackson's Quicksilver. Singularly enough, the county of Norfolk—that home of the Hackney horse—has never owned a champion at any of the Society's Shows until the year 1892, when M.P. won. This colt, whose great point is unquestionably the exceptionally fine quality which he possesses, is a son of Mr. Henry Moore's first champion Candidate, now the property of Mr. Burdett-Coutts, but who won his great Islington triumph when shown by his owner and breeder, Mr. Moore. In addition to Vigorous, East Anglia also possessed a notable stallion in Mr. C. E. Cooke's chestnut Cadet, a horse who, if he had never done anything else, would have gained enduring fame as sire of the famous chestnut filly Pepita. Sir Walter Gilbey's County Member, too, a grand fore-actioned horse and good-looking to boot, stands, as does Reality, on the borders of East Anglia proper, and apart from his big achievements in the show ring will always be remembered as the sire of the champion mare, Nora. Sir Walter Gilbey, too, is happy in the possession of Danegelt, 174 by Denmark, admittedly the most successful Hackney sire of the age, and a horse that would have done well in the show ring had he been more frequently exhibited than he was. Having mentioned the names of the above horses as illustrations of the truth of the old saying that like breeds like, sufficient attention has been drawn to the advisability of breeding from good-looking sires in preference to plain ones, so long as the blood is right, and the value of the Hackney as a cross for other breeds may now be considered.

It may be rank heresy to express such an opinion, but one cannot help expressing the conviction, that Hackney blood, if properly and intelligently made use of, would be extremely likely to improve the breed of modern hunters. Earl Spencer, at all events, for one, appears to have this view, and his experiments will undoubtedly be followed with interest by breeders. Why the Hackney should be decried as a hunter sire it is hard to see, unless the fear exists in some minds that a trotter would be the only result of such a cross, but the truth of such apprehensions, flattering though they be to the prepotency of the Hackney, are unquestionably far fetched. On the other hand, he can instil temper, shoulders, back, loin, and quarters into his stock, and these qualities are assuredly desirable acquisitions in the case of a hunter.

Finally, two short words of advice to those who have never kept a Hackney for general purposes—" Try one." He is strong in constitution, and the best ride or drive horse in existence, providing that he is properly done by; therefore, whilst once more counselling a would-be buyer to be extra firm on the question of pedigree, in order that when he asks for a Hackney he may get one, the advice is, go in and pay a good price for a good horse. You will never regret your bargain.

CHAPTER III.

CLEVELAND BAYS AND YORKSHIRE COACH HORSES.

THE Cleveland Bay is one of the oldest breeds of English horses, though the name by which it is now known is of comparatively recent origin. The Chapman or pack horse, which is the older name, and by which, until quite recently, the breed was known in the more remote of the Yorkshire dales, which became its principal home, points not only to the antiquity of the breed, but to its great utility during the earlier years of our country's history. The breed flourished exceedingly when the roads in more remote parts of the kingdom were little better than tracks, and when the business of the country was principally carried on by its means. Active and powerful, Cleveland Bays were then used as working horses on the farm as well as to convey corn and other marketable produce to the various towns, and their masters to " kirk, or market, feast or fair." In hilly Devonshire and in the north of Yorkshire they flourished the longest, and now it is in North Yorkshire and the neighbouring districts alone that there is anything to be found approaching in type to the ancient pack or Chapman horse. Many are the theories which have been promulgated concerning the origin of the Cleveland Bay, for it will be found more convenient to adopt the modern name. Amongst others which have been received and transmitted in a remarkable manner is that which I believe originated with Professor Low. If he did not originate it, it is one which received his sanction, and with which his name is inseparably connected. This theory is that the Cleveland Bay is the result of an elaborate system of crossing between the thoroughbred stallion and the cart mare. It is singular that such a theory should have received a moment's credence, either from practical breeders or from scientific men. In the first place, the shape of the Cleveland Bay points to the improbability of such a descent, the length, and particularly the long level quarter being such as is never found in any descendant of the cart horse that I have seen. If Dr. Knox is correct in assuming that man "cannot even produce and maintain a new and permanent variety of a barn door fowl, of a pheasant, of a sheep or of a horse," this theory of a cross between a thoroughbred stallion and a carting mare falls to the ground at once, and though I am bound to admit that Dr. Knox' seems a somewhat sweeping assertion, yet undoubtedly physiology points out to us that cross-bred animals do not breed regularly to type, and that the produce of such animals is sure in the third or fourth generations, if not earlier, to revert to the type of one of the original parents. Singularly enough, too, this atavism generally shows the worst instead of the best characteristics of the original parent. It may be said that the fact of the Yorkshire Coach Horse breeding with such trueness to type and character is a practical refutation of this proposition. But if the circumstances are examined they would seem to be a strong confirmation of it. In many respects, from an anatomical point of view, there is a great similarity between what is known as the thoroughbred horse and the Cleveland Bay. There is the same clean flat bone and well-defined sinew, a similar density of bone, such as is possessed by no other breed of horses save the thoroughbred and Arab; the same level quarters and elegant appearance, and the same liberty of action, and though in a different degree, the same hardy constitution and staying power. The fact that the Yorkshire Coach Horse breeds true to type and colour, tends to prove in the main that there is some similarity of type between the thoroughbred and the Cleveland Bay.

The most feasible theory as to the origin of the Cleveland Bay breed seems to be that it has been produced, by a system of natural selection, from the original breed of horses found in the southern part of the island of Great Britain. Probably, nay possibly, an Eastern cross may have found its way into the breed at a very early age. Historical probability and the experiences of a later generation are all in favour of such a theory. That there was a powerful and active breed of horses in the island at the time of the Roman invasion is an undoubted fact. The heavy war chariots with which the Iceni discomfited the veteran soldiers of Julius Cæsar must have been horsed with animals possessing size, strength, and action in a marked degree, possessing, in a word, all those attributes which are comprised in the modern phrase "quality." Cæsar, we are told, was so impressed with the good qualities of the British horses that he took some with him to Rome, and we have the authority of the coinage of King Cunobelin that horses were much valued in Britain, and the portraits of horses found on the coins of his reign, though of course somewhat rude in execution, point to the existence of a breed of great excellence, not dissimilar in many respects to the Cleveland Bay.

It has been suggested, and with some show of reason, that the blood of Eastern horses had been imported into Britain before the galleys of Cæsar set sail from the shores of Gaul. It is well-known that the Phœnicians carried on a considerable trade with the inhabitants of the southern and south-western portions of the island, and it was far from improbable that they would bring over some of their native horses for purposes of sale or barter, especially when they found that the Britons were of a horsey tendency, and had no objection to an honest deal. The similarity in type which existed between the Cleveland Bay and the Devonshire pack horse has been cited in confirmation of the theory that the former is a descendant of the ancient British horse. In Yorkshire and in Devonshire have survived the type of the ancient horse, because in Yorkshire and in Devonshire this type was more strongly pronounced than in any other part of the country, and because in Yorkshire and in Devonshire were bred in the largest numbers the animals which modernised, if I may use the term, the type of the national horse. The Eastern blood which the Phœnicians would be the likeliest to import to Britain would naturally be the Eastern horse of which they possessed the greatest numbers, that is, the Barb. Then we know from historical records, that a legion of the Crispinian horse was stationed at Danum (Doncaster) during the Roman occupation of Britain, and it is equally a matter of history that the Crispinian legion was mounted on Barbs, and it almost goes without saying on stallions of that breed. That these would be crossed with the mares of the country may be taken as a matter of course, and the fact that in the south-western hills and moorlands and the north-eastern dales there existed until lately two breeds of horses which were in many respects of the same type and character, points out strongly that the two breeds must have had a similar origin, and seems conclusively to knock on the head the theory of a cross between the thoroughbred and the heavy-bodied, feather-legged, and, comparatively speaking, unwieldy cart horse.

In order to account for one peculiarity in the Cleveland Bay, the black points, a theory has been started that the Scandinavian horse, who still has those black points very strongly accentuated, was responsible for their introduction into the Cleveland Bay breed, and that his introduction when the country was over-run by the Saxons and Danes, though not sufficient to materially alter the type of the native horses, was sufficient to leave a mark upon them which lasted through many generations, and which is now gradually dying out. The theory is a very ingenious one, but it will not hold water for a moment. In the first place, it is in direct opposition to the physiological principle to which I have already drawn attention. The crossing with such a straight-shouldered, undersized, crooked-hocked commoner as the Scandinavian horse must have resulted in lamentable failure. Again, the black points have been seen by Darwin very strongly defined on a dun Devonshire pony. The pony not only had the zebra-like stripes on the legs, and the mark down the back which was so long a leading characteristic of Cleveland Bays, and which is now very rarely to be found, but he also had broad shoulder marks as well. Darwin describes the pony as a fallow dun —that is, between "a cream and a reddish brown which graduates into light bay or light chestnut;" and it is worthy of notice that it was amongst the light bays that the black points of the Cleveland were the most frequently found and the most strongly marked. There are also instances of the black points being found strongly marked on light bay or dun cart horses. Racing men speak of Doncaster's black spots, and these black points which old Cleveland Bay breeders used to value as such an infallible sign of purity of blood would seem to be common, in a greater or less degree, amongst all breeds of horses, and to be in some measure a reversion to the feral horse.

Leaving the region of theory, we come to the fact that the existence of a breed of clean-legged active horses, clear of thoroughbred and carting cross, was acknowledged quite two hundred years ago. Unfortunately, the men in whose hands this valuable breed of horses was principally to be found, did not keep much record of their stock in writing, and it is therefore on oral tradition that we have principally to rely for our early history of the Cleveland Bay horse. I have in my possession a letter which was written some eight years ago by a man who was fast approaching his eightieth year, in which he told me he had heard his great grandfather speak of the breed with enthusiasm, and he claimed to have direct descendants of a breed, the taproot of which had been in the possession of ancestors still more remote. I have also heard the late Mr. Lumley Hodgson—than whom no one was better able to form an opinion, and who was perhaps possessed of as much horse lore as any man of his time—say that when in the early years of the century he went buying young horses in the East Moor Dales, old men used to tell him of the bright bays, "clear of blood and black," that were recognised as a pure breed by their forefathers before the days of the Darley Arabian and the Godolphin Barb. The following passage from Tuke's "General View of the Agriculture of the North Riding" is of interest as bearing on this part of the subject:—"Yorkshire has long been famed for its breed of horses," says he, "and particularly this riding, in almost every part of which numbers are still bred, the prevailing species of which are those adapted for the coach or the saddle. In the northern part of the vale of York the breed has got too light in bone for the use of farmers, by the introduction of too much racing blood; but the most valuable horses for the saddle, and some coach horses, are there bred. In Cleveland the horses are fuller of bone than those last described; they are clean, well made, very strong and active, and are extremely well adapted to the coach and the plough." Tuke goes on to say that in the southern part of the vale of York, the Howardian Hills, Ryedale, and the Marishes, a greater admixture of "black," *i.e.*, carting blood, prevails, but that the district still produces a considerable number of coach horses; whilst the East Moorlands, he assures us, though possessing a hardy and active breed of horses, did not produce many that were big enough to horse a coach. One more quotation from Tuke may prove of interest:—"The horses which are sold for the London market, if for the carriage, are chiefly bay geldings, with but little white on their legs and faces; those which have much white, with chestnut, roan, and other unusually coloured horses and mares, generally do not bear an equal price in the London market, but, with other slight and undersized horses, are more sought after by foreigners, and eagerly purchased by them for exportation."

Until the earlier years of the eighteenth century, it seems fair to presume that Coach Horses, or Chapman horses, or Cleveland Bays, by whatever name we may call them, were not known as a distinct breed—a surmise which an extract from the Note Book of Sir Walter Calverley, dated January 15th, 1670, goes far to confirm. He relates that when he wanted to use his coach, he horsed it with the lighter "mears" of the breed used on the farm. From the account Sir Walter Calverley gives of the performances and behaviour of these same mares, it seems pretty clear that they were much more active and lively than the heavy draught horses, with Flemish blood in their veins, could possibly be. What was Sir Walter Calverley's custom seems to have been an universal one in the seventeenth century, and if these lighter mares were consistently bred from, as they undoubtedly would be, their lightness and activity would gradually develop and increase, until they became the chief characteristics of a practically new breed.

It is worth while for a moment to give some consideration to the history of the polled cattle in connection with the development of "new" breeds. On the authority of Youatt, who was a keen observer, in the year 1750 a proportion of the Galloway cattle had horns, yet within sixty years of that time a horned Galloway was scarcely to be found, and now one is quite unknown, and no breed of domestic animals breeds so true to type. In a kindred breed, too, the Aberdeen-Angus, a striking modification of colour has taken place during the same period. In the middle of the last century, cattle of a dark red colour were found amongst them, although not in such numbers as horned cattle were found amongst the Galloway breed, but the fiat of the pioneers of the breed had gone forth that black was to be the colour, and no animal of any other colour was ever used for breeding purposes. When considering the probable development of the Cleveland Bay from the native horse, the question of spontaneous variation also deserves some consideration. On this point Darwin is very clear. "It is probable that some breeds," he states, " and some peculiarities, such as being hornless, &c., have appeared suddenly owing to what we may call in our ignorance spontaneous variation," and that through selection in breeding, these spontaneous variations have come to possess a powerful hereditary tendency. "It is admitted by all authorities," say the authors of an invaluable work on Polled Cattle, "that while deviations from the original or typical form or race of animals may arise spontaneously, some sort of artificial method or selection in breeding is necessary to impart to those spontaneous and isolated deviations such fixity of character, or strong hereditary power, as would insure their perpetuation."

It seems only reasonable to suppose that the Cleveland Bay may have had its origin in a similar natural selection as that which has developed the Galloway and the Aberdeen-Angus cattle. Such an origin would be far more consonant with the principles of physiology than any elaborate system of crossing, and the prepotency of the Cleveland Bay seems to confirm the fact that the breed has been produced in some such manner as I have suggested. There are writers who, anxious to account for every characteristic and good quality which the breed possesses by deriving it from some other source, maintain that the hardihood of constitution, the courage, and the activity of the Cleveland Bay could only spring from a strong infusion of Eastern or thoroughbred blood. But the native breed of horses was undoubtedly hardy in constitution, very active, and possessed of comparatively a fair turn of speed. It was especially valued by patriotic Englishmen, and so late as 1739 the introduction of Eastern horses into England was bitterly deplored in a curious article to be found in the *Gentleman's Magazine*. The writer, speaking of horse-racing, says:—" The original design of this entertainment was not only for sport but to encourage a good breed of horses for real use, and the Royal Plates are supposed to be given for that purpose, the horses being obliged to carry heavy weights; but alas! how are these intentions perverted; our noble breed of horses is now enervated by an intermixture with Turks, Barbs, and Arabians, just as our modern nobility and gentry are debauched by the effeminate manners of France and Italy." The fact that English horses possessed courage, speed, and endurance, as well as strength, is sufficiently proved by Mr. Joseph Osborne, in his interesting introduction to the "Horse Breeder's Handbook," and therefore it is not necessary to look for the origin of the courage and hardy constitution of the Cleveland Bay to the Arab or the Barb; and indeed stallions of either breed would not be likely to be within the reach of Cleveland Bay breeders.

But that an occasional cross of the thoroughbred was introduced is by no means improbable, though it is remarkable that an instance of its occurrence is not found in any historical record of the breed. It is none the less remarkable that those who tell us that the modern Cleveland Bay is the result of crossing with the thoroughbred and the carting mare, are especially careful to avoid naming the thoroughbred or to give any reliable data about such crossing. Still there can be little doubt that an occasional cross of blood was accidentally or designedly admitted into the breed. In later years rumours to this effect were extant, and in one instance a prize was withheld on this account, but the rumour may be taken for what it was worth; for when the owner of the disqualified animal sued for the prize in the County Court, as eventually he did, those who stated that the so-called Cleveland Bay was by a thoroughbred signally failed to prove their case. The probability is, that if the thoroughbred cross was introduced, it would be in the middle of the last century. Any introduction of alien blood in later years would be sure to be known, and indeed there would have been no object in hiding it, as, until very lately, Coach Horses were more valuable than Cleveland Bays. One thing that points to the introduction of thoroughbred blood at the time I name, is that records state how well some men were carried to hounds by their pure-bred Clevelands in the latter part of the eighteenth and the commencement of the present century; and although hounds did not go quite so fast then as they do now, the country was principally undrained, and consequently a "bit of blood" would be required to get through the deep ground in the style which the old stories tell us these Cleveland Bays did.

Probability points to a thoroughbred named Traveller as having had something to do with imparting fresh quality and courage to the Cleveland Bay. On many of the cards and bills of the older stallions the pedigree is traced back to a certain Old Traveller, and then stops. There is no pedigree of the Old Traveller given, and as is usual with old stallion bills, the language is obscure, and the identification of horses named becomes a matter of difficulty. But the constant recurrence of the name of Old Traveller in the old bills would seem to point out that he was some well-known and highly-appreciated horse; and a thoroughbred horse of that name—a thoroughbred horse that was afterwards to make a great reputation at the stud—did travel in the neighbourhood of Yarm, in the middle of the eighteenth century, serving mares at a nominal fee. And it must be borne in mind that Yarm was in the very heart of the country where Cleveland Bays most flourished. In after years many famous horses were bred within a few miles of what was then one of the most important towns in the Vale of Cleveland, a town whose horse fair, though now decayed and of little importance, was at one time one of the most important in the north—one, moreover, at which Cleveland Bays were to be found in greater numbers than at any other fair, excepting Northallerton. Taking these facts into consideration, together with the fact that a thoroughbred horse was serving mares in the vicinity at a nominal fee—always an important affair in a country district, and more especially so in those days—and it is not difficult to imagine how this Old Traveller might have done much in imparting quality to the produce of the Cleveland Bay mares of the district. Horses of which he was the sire, and of whom no record remains, may have been kept as stallions and transmitted his stoutness and quality to another generation; mares by him may have, and probably they did become, famous as brood mares, and their stock would probably show more than ordinary elegance and style.

It is interesting to pause for a moment and recall this famous old horse. He was bred in 1735 by Mr. Osbaldeston —the grandfather of the well known " squire," who was afterwards to make such a name in the world of sport—and was a bay colt by Partner, dam by Almanzor. Out of the six races in which he took part, and of which a record exists, he won four, and was disqualified for another on account of a cross, and it certainly seems curious that after so successful a career on the Turf he should come to be travelling in a country district at a very nominal fee. Fortunately some of his half-bred stock showed promise, the Duke of Cleveland and Mr. Shafto sent some good mares to him, and he became the sire of Dainty Davy, Squirrel, Lass of the Mill, and other good race horses.

In the middle of the eighteenth century horses of the highest character stood at remarkably low fees; indeed, well into the middle of the present century, classic horses were occasionally to be found whose fees for half-bred mares did not exceed two guineas and a half. It is therefore not unlikely, nay, it is very likely, that occasionally a man might cross some highly prized mare with one of these horses. But it is remarkable that in few of the old pedigrees that I have examined have I ever seen the acknowledgment of a foreign strain of blood. It seems to have been looked upon with a certain degree of suspicion even in those days, and though it may have been of occasional, it was by no means of frequent, occurrence. The traditions of the old breeders who boasted of their long line of horses in whose veins was neither "blood nor black" were in the main respected, and though there might be an occasional use of the thoroughbred, I am inclined to think that except when for the avowed purpose of developing the Coach Horse breed, such crossing was very little resorted to.

In the middle and latter part of the eighteenth century and in the earlier years of the present century, the whole of the agricultural work in the Vale of Cleveland was practically performed by Cleveland Bays. The surface of the country presented a very different aspect to what it does now; there was a larger proportion of grass, and the Cleveland Bay was powerful enough to do all the work of the farm. Indeed, I should very much question if, in the heart of Cleveland, the draught horse in anything approaching his modern type was known at all until the present century was eight or ten years old. With the wars which were the direct result of the French Revolution, the value of wheat and other cereals rose to famine prices, and when oats sold, as they did sell, at 6s. 6d. per bushel, whilst wheat made as much as a guinea, it was not to be wondered at that farmers and landowners equally were eager to grow corn wherever corn could be grown, and that they hastened to convert into tillage much of the good grass land in the Vale of Cleveland. Nor were they content with turning their pasture into tillage. Bleak and apparently inaccessible places—places more adapted for the growth of larches, or even Scottish firs, than corn—were broken up and sown with wheat. Indeed, the memory of one of these rash enterprises is preserved in the Ordnance map by its name of Bold Venture. A sweeping change like this, as a matter of course, brought other changes in its track. Farmers who were resident in the neighbourhood of Stokesley used to take their corn to Thirsk market, a distance of some twenty miles over not the best of roads, and then they began to fancy that they required a heavier and more powerful horse. The land which had been converted into arable was also found to be a strong clay, and on this account again the farmers thought that they required a heavier breed of horses. So they crossed their fine Cleveland mares with such cart horses as they could get, with a result that was disastrous, and which indeed nearly

Cleveland Bay Stallion, Sultan 667.

proved to be fatal to the existence of the Cleveland Bay as a distinct breed. Another circumstance which took place about the same time also did much harm. This was nothing more than a change of fashion. It became the rage to drive in curricles big upstanding horses, approaching and sometimes exceeding seventeen hands in height; and to meet the prevailing fashion fine Cleveland mares were mated with leggy, flash-topped thoroughbred horses to a very considerable extent. These causes, coming together at the same time as they did, nearly put an end to the pure bred Cleveland Bay.

This crossing of the leggy thoroughbred with the Cleveland mares was also, in the opinion of the late Mr. Lumley Hodgson, responsible for another evil; and he traced the great increase of roaring to this cause. Animals bred this way with their fine "rainbow necks" were, of course, peculiarly liable to this disease, and as many of them were kept as sires, and, moreover, were largely used, it is not improbable that a large proportion of the roaring which is found amongst Coach Horses and Cleveland Bays in the present day may have had its origin in the fashion for over-sized curricle horses which sprung up in the days of the Regency. I have heard it stated that previous to the time when this "crossing" took place, roaring and its concomitant diseases were unknown amongst Cleveland Bays. I give the tradition for what it is worth, and cannot say that I place much credence in it. There is no doubt that many horses which were considered sound a hundred years ago would not pass the stricter examinations of the present day successfully, and perhaps this may account for the tradition.

Referring to the decadence into which the breed had fallen from causes enumerated above, the *Farmers' Magazine* for 1823 says, speaking of the Cleveland Bay:—" It was the basis of the old London coach horse, when heavier cattle were used for those conveyances; and after the fashion became to adopt a lighter horse for carriages, this valuable breed was allowed to become almost extinct, till their excellence for agricultural purposes was noticed by some practical farmers in the north of England, who for several years back have been exerting themselves to revive the breed."

It is a subject for deep regret that more is not recorded of the efforts of these astute and far-seeing gentlemen, to whom subsequent generations are so much indebted. Unfortunately they lived in an age when little importance was attached to the recording of facts connected with stock breeding and agriculture. Newspapers were scarce; what agricultural literature there was, was published in London, then quite out of touch with North Yorkshire farmers; and it was impossible for the local newspapers, which were in existence in those days, to devote any attention to the subject. Indeed, those who were most interested did not consider it necessary to make any notes themselves, and pedigree seems to have been thought of so little importance, that even the names of the sires of Mr. Masterman's famous horses, Skyrocket, Summercock, and Forester, are unrecorded; neither is it possible to more than approximate the date at which they flourished. Tommy Masterman, a Cleveland farmer, was one of those who took an active part in the revival of the Cleveland Bay breed, and his exertions in this direction were so highly appreciated that he was presented with a testimonial by his friends and neighbours. This took the form of a silver cup, value £10, which was presented in 1820, and the inscription on which states that it was given in recognition of the service which Mr. Masterman had rendered in keeping first-class sires. I have also heard it asserted that the name of Skyrocket was mentioned in the inscription, and from common report Skyrocket seems to have been the best of Mr. Masterman's horses. Some years ago I made every effort to trace the owner of this cup, but without avail. A portrait of Skyrocket was once extant, but I have been unable to discover it; and I am informed that the old signboard of the public house at Nunthorpe, in Cleveland, was a copy of this picture. This public house is now done away with, the sign has long been destroyed, and the only portrait of Skyrocket which is in existence is engraved on a glass in the possession of my brother, the said glass being one of half-a-dozen which Mr. Masterman had engraved. The portrait, which can scarcely be expected to be an accurate likeness of the horse, shows him to have been of exceptional quality, in many respects not unlike a thoroughbred horse. He is also represented as short of substance, and it is highly probable that the artist had drawn considerably on his imagination, as is frequently the case with portraits of our older horses. The horse is nicely turned, and has a remarkably fine outline.

Another gentleman who did much for the breed was the late Mr. John Richardson, of Langbarough Hall, near Great Ayton. Mr. Richardson was an enthusiastic breeder, and spent much time and money in endeavouring to raise the standard of Cleveland Bays. Strange to say, although he bred many stallions, he does not seem to have had any of any great merit with the exception of Drainer, from whom descend many famous horses and mares, amongst the latter Tommy Peart's Darling. But Mr. Richardson's mares achieved a wide celebrity, and perhaps no man did more in his day than he to bring the breed into prominence.

About the same time Cleveland Bays were taken into both north and south, with a view to the improvement of other breeds, and the results alike in Scotland and the south-western counties were highly satisfactory. Indeed the breed was recognised and valued all over the country, and those practical farmers in the north, of whom the *Farmers' Magazine* spoke, were amply rewarded for their enterprise and energy.

From the time of which I have been speaking, up to the year 1867, Cleveland Bays flourished exceedingly, and there seemed to be no likelihood of the breed again falling into desuetude or decay. Mr. Hansill, Mr. Thomas Peart, Mr. York, of Worsall, Mr. John Smith, of Long Newton, the late Mr. Robinson Watson, Mr. John Robinson, of Hutton Rudby, and others, bred largely, and were the owners of valuable animals. But in 1867 came a reaction, as a reaction had come earlier in the century, and singularly enough, the heavy cart horse had something to do with the reaction which set in twenty-five years ago. The Cleveland iron trade had increased with leaps and bounds, and in other seven years was to reach its zenith. The demand for heavy horses, adapted for drawing heavy loads on the roads and in the mines, became larger and larger with each succeeding year, and prices for them were abnormal. The great improvement in the breeding of cart horses, which began to manifest itself about this time also, gave an additional impetus to the breeding of heavy cart horses, and Cleveland Bays were gradually more and more neglected. The foreigners came in, and bought what they could of the best, and the men who kept their mares, bred hunters from them, and crossed them out of recognition. Cleveland Bay classes ceased to fill, and finally were, with one or two exceptions, dropped out of the prize schedules altogether, and it seemed as if the breed must become extinct.

But when things are at the worst they not infrequently begin to mend. At any rate that was the case with the Cleveland Bays. At the very time when Englishmen generally looked upon the breed with feelings akin to contempt, there was a growing idea in the United States that this was the class of horse that was wanted in their country, and just when the fortunes of this valuable breed were at their lowest ebb, when only after a warm discussion and a close division was it possible to retain even a couple of Cleveland Bay classes in the representative Society of the district, there was beginning—in a very modest way, it is true—that trade with the United States which has since increased to large dimensions.

In the dales about Whitby, and running down to the east coast of the North Riding of Yorkshire, the Cleveland Bay had been tenaciously preserved. The farmers there were proud of their horses, the breed of which had been in their families for generations, and they never took kindly to the

Clydesdale, or Shire horse, whose "feather" they looked on—and still look on—with dislike. It is to this tenacity of purpose that we owe the existence of the Cleveland Bay as a distinct breed at the present day. Mr. Hindson, of Ugthorpe, who has had a capital strain of Cleveland Bay horses all his life, kept some good stallions, and his example was followed by Mr. John Welford, who, like Mr. Hindson, has always owned a stud of high class Clevelands. Then the Right Hon. James Lowther came to the rescue, and not only purchased Fidius Dius at the Guisbrough Park sale, but set to work to get some good mares together. Amongst others he purchased the descendants of the mares that had been bought for Earl Fitzwilliam by the late Admiral Chaloner—a purchase which has proved a distinct gain, not only to Mr. Lowther's stud, but to the country, for the best blood in the country ran in the veins of the mares which hailed from Coollatin. They were by Brilliant, an elder brother of Captain Cook, who was very successful both in the show ring and at the stud, and a son of Harry York's Wonderful Lad and Mr. Peart's famous mare Darling. So that when the time came that Cleveland Bays were wanted, there were plenty to be found. Not that they were there in any great numbers at first. Men were not so careful about breeding them as they are now, and many of the mares were mated with thoroughbred horses, with the object of breeding hunters; but there were quite plenty of mares, and stallions too, to form the nucleus of a breed.

The revival of the general interest in Cleveland Bays may be said to date from 1883. In the previous year it had taken all Mr. T. Parrington's influence to get a class for Cleveland Bay mares inserted in the prize schedule of the York Meeting of the Royal Agricultural Society, and only one mare was exhibited. This was Mr. W. D. Petch's Fanny, a mare that he sold to Mr. A. E. Pease, and that subsequently found a home in the Brookfield stud.

But if Fanny was the only exhibit, she was a very useful specimen of the breed, and the judges considered her worthy of the first prize; "only one entry, very good," being their note on the class in the official report. She was then eight years old, was full of bloom, and combined quality and substance in a manner that was unknown to many visitors to the Royal Show. No wonder, then, that she was the object of considerable attention.

In this same year, too, the rivalry between Mr. Codling's Blossom and Mr. Welford's Madam began to attract the notice of show-yard visitors. Both were Cleveland Bays of good pedigree, both were fine movers, and had remarkable quality; so when at the latter end of the year 1883, attention began to be called to the merits of the old breed, the public mind—in the north of England, at any rate—was to a certain extent educated on the subject.

The formation of the Cleveland Bay Horse Society, in January of the following year, consolidated and gave expression to that vague feeling which had been growing, that the Cleveland Bay was a breed to be cultivated.

Since then the breed has increased in numbers in a satisfactory manner. All over the country gentlemen have begun to breed high class Clevelands. In Northumberland and in Hampshire, in Essex, and in the neighbourhood of London good studs are to be found. But perhaps even more marked than the increase in numbers is the improvement in quality which has taken place. It has been a frequent remark amongst exhibitors of late that animals have no chance of winning now, that eight or ten years ago would have won readily enough, and good judges have stated that the average merit of the young animals in the Cleveland Bay classes was higher than they ever remembered to have seen, and their memory extends over many years. It is satisfactory to note that the improvement which has been effected in our other breeds of horses has taken place in the Cleveland Bay breed in quite as conspicuous a manner, and it may now be fairly hoped that it will never again run the narrow risks of becoming extinct that it has done in the past.

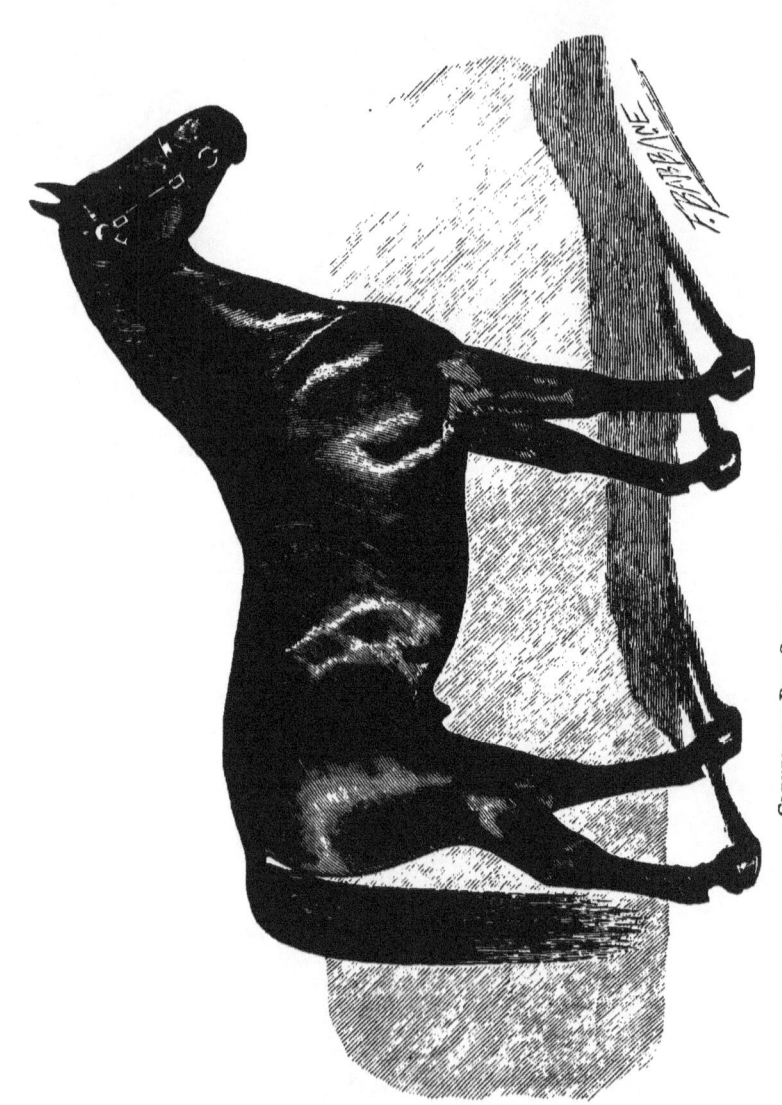

Cleveland Bay Stallion, Master Frederick 992.
Owned by Mr. James F. Crowther.

The Cleveland Bay is the embodiment of quality and substance combined. In height he stands from 16.1 to 16.2½ or 16.3, rarely exceeding the latter or falling short of the former height. He stands on a short leg, his shoulders slope well, his back and loins are strong, and his quarters are long, level and muscular. Indeed, the elegance of the quarters and the set on of the tail are amongst the distinguishing features of the breed. His head is rather plain, but it is a good lean head, and is well carried. The bone is flat and clean, the legs devoid of hair or nearly so, and the sinew is clearly defined. In many respects the shank bones of the Cleveland Bay, the thoroughbred, and the Arab resemble each other. The shape and density of the bone, and the clearness with which the sinew stands out are particulars in which the resemblance may be noticed, and although it would perhaps be too much to say that in density of bone the Cleveland Bay equals the thoroughbred or the Arab, it can be asserted with confidence that he stands far in advance of any other breed.

The action of the Cleveland Bay is one of his strong points. There is no "snap of the knee" such as is seen in the Hackney, or at any rate there is very little, and what little there is, is in all probability the result of training. But the shoulder action is excellent both in the walk and trot, in the latter pace not being unlike that of the thoroughbred. The hocks are well flexed and got well underneath the body, and as the Cleveland Bay covers a lot of ground it is easy to see that he can travel at a good pace. But it is necessary that he should possess substance as well as style, for one of the most useful purposes to which he can be put is farm work. The measurements of a Cleveland stallion which Mr. J. B. Lloyd took into Gloucestershire in 1827, gives a good idea of a typical specimen. "When old Cleveland," says Mr. Lloyd, "was at his full size he measured 16 h. 1 in. high, 9¾ in. round the pasterns, 10 in. round below the knee, 21 in. round the arm, 15⅝ in. round the knee, and 6 ft. 10 in round the girth. When measured he was in good condition, but not what you would call full of flesh; his legs as clean as a race-horse."

I fancy Cleveland must have been an exceptionally powerful horse, and as he was purchased with the avowed object of improving the breed of agricultural horses in Gloucestershire, it is probable that this was the case. At any rate, such a measurement as 10 inches below the knee, is, I should say, extremely rare in the present day.

As the name of the breed would seem to imply, the colour is bay, the legs a good black, and although a small white star, or a few white hairs on a hind heel are not an *infallible* sign of alien blood, they are regarded as an eyesore; and unless they are very small indeed, so small as to be scarcely visible, the sale of an animal possessing them is very much affected thereby. The black points are of rare occurrence now-a-days. They consist of black, zebra-like stripes on the arms and thighs, just above the knees and hocks. Occasionally there is a black stripe down the back or a black or dark spot or two on the quarter, the black stripe being generally found on horses of a light bay colour, whilst the stripes and marks on the quarter have been more generally associated with animals of a darker colour.

Some misapprehension seems to exist respecting the colour. It is said by some that the bright golden bay is the only colour which is admissible and that a darker colour implies the existence of alien blood, but such an idea is manifestly erroneous. From the golden bay, and even the fawny bay, to the dark bay the difference is only one of degree, and it is difficult to see why a prejudice should exist against a dark bay. Darwin's conclusions on the subject of colour are so clear, and seem to have so much bearing on the *quæstio vexata* of light and dark bay, and the dappling of which so much is made in some quarters, that I quote them:—

"Horses occasionally exhibit a tendency to become striped over a large part of their bodies, and as we know that stripes readily pass into spots and cloudy marks in the varieties of the domestic cat and several feline species—even the cubs of the uniformly coloured lion being spotted with dark marks on a cloudy ground—we may suspect that the dappling of the horse, which has been noticed by some authors with surprise, is a modification or vestige of a tendency to become striped."*

This would seem effectually to dispose of the statement made in many quarters that a dark or dappled bay is a sign of alien blood, which, coming from men who insist strongly on the "black points" as a sign of exceptional purity of breed, is surely inconsistent. At the same time it must be borne in mind that although the darker coloured horses may be well-bred ones, and trace their descent for many generations through famous horses and mares, the bright bay is much to be preferred, and for many generations the opinion has prevailed that this is the colour *par excellence*.†

In concluding this description of the Cleveland Bay, especial attention should be called to his hardihood of constitution. No pampering is required; he will do his share of work for many years, and perhaps few breeds of horses are so noted for longevity or are such prolific breeders. If it were necessary to describe the Cleveland Bay in a word, it would be done by designating him the general utility horse, a name by which I believe he was at one time known in some parts of the United States. As a matter of fact no work that he can be put to comes wrong to him, except, of course, fast work, for which he is not adapted. In the plough, on light or medium land, he will work the heavier Shire or Clydesdale to a standstill, his superior activity giving him the pull; and he will always come home with his head and tail up. Even on

---

* "Animals and Plants under Domestication," vol. i., p. 56.

† As bearing on the question of light and dark colours in horses, it may be of interest to mention an instance of change of colour which happened to a horse of my own. He is a hunter, and is now (1894) six years old. In the spring of 1893 he was a red chestnut, approaching to sandy. He is now a dark liver chestnut.—W. S. D.

strong clays, Cleveland Bays have been known to hold their own, as the following anecdote goes to prove. In the early part of the century a large farmer moved from the neighbourhood of Darlington to Northumberland, and as a matter of course, he took his Cleveland Bay horses with him. Then, as now, Northumberland was the home of the heavy and powerful draught horse, and the Northumbrians justly prided themselves on the excellent breed of horses they possessed. It was only to be expected that they would hold the comparatively light horses of their new neighbour in derision, and they were very free in their criticism of his teams. But the new comer was not to be chaffed with impunity, and one market night he was stung into challenging the country to a ploughing match. His challenge was speedily accepted, and the terms, which were very simple, were arranged without a hitch. They were as follows. Each party was to produce a pair of horses on the following Monday morning, and they were to plough from Monday morning till Saturday night, the pair which had ploughed the most land in the time to be declared the winner, the stakes being £50 a side. On Monday morning they commenced to plough accordingly, but before Wednesday night the heavier horses had had quite enough of it, and the Cleveland Bays were declared the winners. Though I should scarcely be inclined to recommend the modern Cleveland Bay for heavy farm work on the strongest clays, yet there is no farm on which an active Cleveland Bay mare cannot be made to pay her way, and pay her way well. The countless jobs which require activity rather than massive strength, and which it would be tedious to enumerate, are much better done by an animal of this type than by a Clydesdale or Shire horse. In olden times Cleveland Bays have been used on occasion as hunters, but with the pace hounds run now-a-days that occupation for them is out of the question. They are still, however, occasionally used as carriage or dog-cart horses, and if not so showy as the Hackney or so stylish as the blood Coach Horse, they have a good appearance and get over the ground in a creditable fashion.

It has been said that there is no breed of horses—leaving, of course, the racehorse out of the question—which a cross with the Cleveland Bay will not improve, and the Cleveland Bay has been likened to the Shorthorn and the Leicester sheep, as the best foundation on which to commence crossing. Indeed, it is the value of the Cleveland Bay brood mare in this direction which is a constant source of danger to the breed. Men buy well-bred mares, cross them with the thoroughbred, or in some instances with the Hackney; they are satisfied with the result of their enterprise, and keep breeding on the same lines, getting valuable horses such as they require, but running a great risk of "killing the goose that lays the golden egg." A pure-bred foal or two should always be taken from well-bred mares, for it is little short of a national loss when any famous strain of blood, to whatever breed it may belong, becomes extinct. In breeding Cleveland Bays great care should be taken in selecting mares and sires with good shoulders. In this respect there can be little doubt that the breed had to a considerable extent deteriorated before the revival of interest in it which took place about ten years ago. Yet it is a point, the importance of which it is impossible to overrate. There are some people I know who hold that a nicely sloping shoulder is not an essential for a horse whose work has to be done in harness and at the draught, but with this opinion I cannot agree. A well-placed shoulder not only adds greatly to the general symmetry of a horse, but it is a distinct advantage to him in doing his work, whatever that work may be. If a horse's shoulder is improperly placed there must be an undue strain on the forelegs, so by all means be particular about the shoulders of both sire and dam. Then there is a special reason for being particular about the shoulders of Cleveland Bays. Cleveland Bay mares are frequently used to breed weight-carrying hunters from, and good shoulders are an important factor in carrying a heavy man.

About the breeding of hunters from a Cleveland Bay mare, great difference of opinion exists. Some assert that animals bred in this way are soft and useless as hunters. That is certainly at variance with the experience of many well-known hard riders in Yorkshire, men who can hold their own in any country. The late Lord Middleton had a famous mare named Magic, who was the daughter of a Cleveland Bay, and some descendants of hers are still to be found in the Birdsall stables, and right good hunters they are. If it is desired to breed hunters from a Cleveland Bay foundation, in the first place a short-legged wide mare with good shoulders and back should be selected. A mare answering this description can be found with a little trouble. Then comes the more difficult task of selecting a suitable stallion with which to mate her. The prevailing partiality—I had almost said craze—for a big horse is to be carefully avoided. Neither do I consider that the bone measurement is of paramount importance. The sire I should choose to cross with Cleveland mares should certainly not exceed 15.3, and I should like him no worse if he did not exceed 15.2. Quality would be the great thing required. His head and neck must be well set on, and above everything his shoulders must be well placed and muscular, and his back loins and quarters powerful. The shape and quality of the bone would be considered rather than its size, and his action would be also of more importance in my eye than his capacity to carry weight according to the recognised standard. Such a horse as I have endeavoured to describe was Perion, who was perhaps the sire of more good hunters than any horse of his generation.

It has been considered by men of experience that the second cross from the Cleveland mare produced the best hunters, and there can be no doubt that they have more quality, and amongst horses bred this way are to be found the best-looking and hardiest animals, of course, always excepting thoroughbreds. In mating a mare by a thoroughbred sire from a Cleveland mare, rather a different stamp of sire may be required. In the first place, size and substance should receive more consideration. But every pains must be taken to avoid using a horse with long cannon bones or weak pasterns, however good he may be in any other respect. Special care should also be taken to avoid a horse at all light in the loin, or with a tendency to be deficient in his back ribs.

Perhaps a better example of the successful management of a Cleveland brood mare would be difficult to find than that of Mr. Thomas Peart's famous old mare, Darling, a great show-yard celebrity during the latter half of the fifties and the first half of the sixties. Peart's Darling bred fifteen foals, of which six were stallions of some repute. Two of these were named Brilliant, one of them the sire of Sportsman and other good horses, whilst from his brother descend many mares of exceptional excellence. Master Thomas was another that did good service in Belgium, and Lord of the Manor who was exported to India by Mr. George Holmes, the well-known Beverley veterinary surgeon, was considered by him to be one of the best horses of the breed he ever saw. Captain Cook and Rosebery were also horses of great excellence that were very successful both in the show yard and at the stud, but the first foal Darling had was to a thoroughbred horse, and it came about in this way: she was sent to Wonderful Lad when a two year old, but as she missed to him Mr. Peart had her covered by Perion. The following year she bred a filly foal which ultimately became the property of Mr. H. W. Thomas, of Pinchinthorpe. That gentleman sent her to Newport, and to him she bred a famous horse that was purchased by Mr. John Harvey, the Master of the Durham County Hounds, and that earned a well-deserved reputation as a hunter, being fast, a good stayer, and of extraordinary constitution.

The Yorkshire Coach Horse.

The Yorkshire Coach Horse owes his origin, according to the late Mr. Lumley Hodgson, to the fashion for driving big upstanding horses, reaching up to 17h. 2ins., in curricles in the early part of the century. To what he termed this pernicious fashion, Mr. Hodgson attributed in a great measure the decadence in the Cleveland Bay breed which took place early in the century, and about the time that these big, flash, half-bred horses came to be used as sires. Continual breeding from these half-bred horses, which were Mr. Hodgson's pet aversion, has, however, eventually developed a type of horse which breeds very true both as to colour, conformation and general characteristics. There is a tendency in the Yorkshire Coach Horse to a loss of substance. Quality is maintained and even improved upon, but the general tendency is to a loss of width and bone. This, however, is now very much checked owing to the action of the Yorkshire Coach Horse Society. Previous to the establishment of that Society in 1886, anything was recognised as a Coach Horse that at all conformed to the type, and a horse with two or three crosses of thoroughbred blood was not infrequently used as a Coaching stallion, whilst half-bred horses—horses that is with a direct cross of thoroughbred blood—were quite commonly used. It is easy to see that such a method of breeding, if indeed method it could be called, must tend to a loss of substance; and frequent recourse had to be had to Cleveland Bays to correct this very serious defect. The favourite plan was to put a Coaching mare of fine quality to a Cleveland stallion with action and substance; and the result of such a cross was generally, nay, almost always, satisfactory. But since the establishment of the Coach Horse Society the Yorkshire Coach Horse has been placed upon a very different footing. For a time the Society recognised the horses with a thoroughbred cross, but soon the question of type, and loss of size forced itself on the attention of the Council, and the lines of admission to the Stud

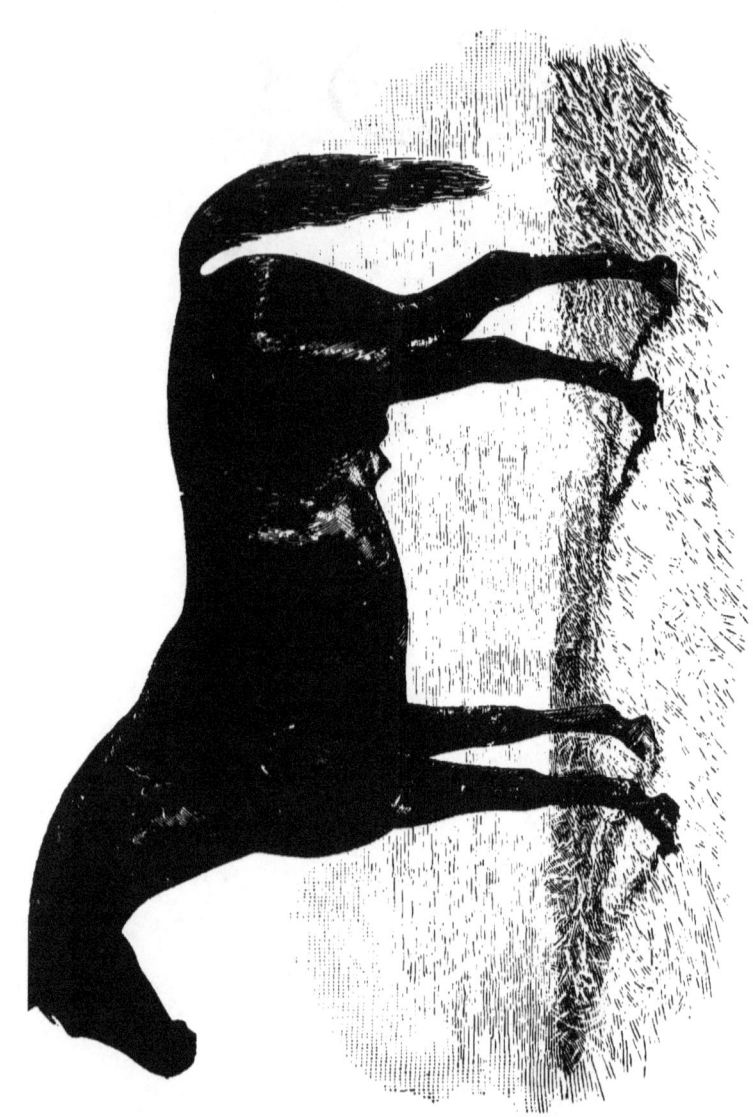

YORKSHIRE COACH HORSE, PRINCE OF WALES 371.
Sire of many Prize Winners.
The Property of Mr. George Burton.

Book were drawn much closer, and nothing with the direct thoroughbred cross was admitted.* The pure bred Cleveland Bay is however recognised as a Coach Horse, probably because of the fact to which I have already alluded, viz., that the best Coach Horses happen to be a cross between the Yorkshire Coach Horse and the Cleveland Bay. How far such a policy is expedient is a question which I do not propose to enter into here. I may say, however, that it certainly does appear anomalous that an animal can be entered in *two* stud books as pure bred. The similarity in many physiological respects of the breeds which have been crossed to produce the Yorkshire Coach Horse does, however, seem to give colour to what, in the case of any other breeds, would appear to be little else than meaningless and arbitrary.

The Coach Horse, though by no means boasting of so ancient an origin as the Cleveland Bay, yet has a claim to respectable antiquity, and is much older than many of our breeds of domestic animals. For fully a hundred years he has had a recognised existence, and prizes were given for him in Howdenshire as early as 1805. It is somewhat curious that on some of the old stallion cards the term "Cleveland Coach Horse" is used, and that these old horses, notably Victory and Volunteer, are claimed both by Cleveland Bay and by Coach Horse breeders as the tap roots of famous strains of horses. As bearing on the question of the name of the breed it is also worthy of remark that the "Druid," speaking of Mr. Jolly, of Acomb, and the trade he did with the Indian Government, refers to his Coach Horses as Howdenshire Clevelands.

I am inclined to believe that a good many of the earlier Coaching stallions were cock-tails. For instance, looking through the first volume of the Yorkshire Coach Horse Stud

---

* Since this was written the Yorkshire Coach Horse Society has again admitted a single thoroughbred cross; a policy which some admirers of the breed strenuously opposed as reactionary, and as calculated to cause harm to the breed eventually.

Book we find many pedigrees like the following. Paulinus, sire Necromancer (S.B), dam by Woldsman (S.B.), granddam by Screveton (S.B.), great granddam by Grog (S.B.). Now, although all the horses whose names are mentioned in this pedigree are to be found in the Stud Book Paulinus is conspicuous by his absence, so the presumption is that his fourth dam would be a Cleveland Bay.

The difficulty of tracing the history of the earlier Coach Horses is also considerably enhanced by so many of them being named after thoroughbred horses, and, indeed, notwithstanding the fact that the breed is of much more recent date than the Cleveland Bay, its early history is enveloped in quite as much obscurity. Records do not seem to have been carefully kept, and there appears to be considerable confusion respecting some of the older pedigrees. For many years Coach Horses contained a very large admixture of the thoroughbred, for example, Quintessence who was foaled in 1830 had five top crosses of the thoroughbred, and upwards of thirty years later, Prince Arthur, the winner of the first prize at the Yorkshire Show, had two top crosses of the thoroughbred.

As is the case with Cleveland Bays the Yorkshire Coach Horses are principally bred by farmers who keep two or three mares, seldom more than half-a-dozen, the heavier of which they work on the light land. Large studs are the exception, but in the neighbourhood of York and Selby, as well as in Howdenshire, there are a few men who breed and graze Coach Horses rather extensively. These gentlemen seldom attempt to cross their Coaching mares with thoroughbred horses unless it is with the object of breeding carriage horses for the London market, a very lucrative branch of the business of horse breeding, and one which obtains largely in the East Riding. It must not be imagined, however, that the thoroughbred sire is generally resorted to when the breeding of London carriage horses is the object aimed at. Coaching stallions are most frequently used, as there is then the chance of breeding a stallion. Cleveland Bay stallions are also sometimes resorted to in the case of very light or undersized mares. A different type of thoroughbred stallion should be used for a Coaching mare than that recommended for a Cleveland mare. More size and length are desirable; the shoulders should of course be good, but the back and loins are not of so much importance, whilst it is of importance that the quarters should be long and level, and that the tail should not droop, but should be well set on and well carried. Indeed, the thoroughbred horse selected should possess as much as possible that elegance of the quarter for which the Cleveland Bay and the Yorkshire Coach Horse are alike remarkable.

The Yorkshire Coach Horse may be described as a Cleveland Bay with more quality, *i.e.*, more of the thoroughbred quality. His head is smaller than the Cleveland Bay, and more blood like, and the crest is more arched. The Coach Horse is also narrower, and has less bone. Indeed, some of the lighter horses show a great deal of the thoroughbred character. His action is good, and perhaps has a little more style about it than the Cleveland Bay, though this may in a measure be the result of training. Coach Horse breeders are as particular about colour as are their Cleveland Bay friends, and though they admit every shade of bay and brown, they will not look upon any other colour, and carefully avoid using a stallion conspicuously marked with white.

The general management which answers for the Cleveland Bay will answer equally well for the Coach Horse. It is true that Coaching mares are not so well adapted for farm work as are the more powerful Clevelands, but on a light land farm there are many jobs for which their activity is especially adapted, and they generally are looked upon to earn their keep as well as breed a foal.

Since the formation of the Cleveland Bay and Yorkshire Coach Horse Societies, both Cleveland Bays and Coach Horses have been more carefully bred, and the difference between the two breeds has been more fully recognised by the general public. This was greatly to be desired in the interests of both breeds, and it is a source of satisfaction to all concerned that the Yorkshire Agricultural Society has at last recognised that the breeds do exist as distinctly as they did in the earlier years of the century, and that that recognition took a liberal form. The Royal Agricultural Society have also a similar recognition of the two breeds under consideration. It would be little short of a national calamity were the two breeds to be merged in one, viz., in that of the Coach Horse. The loss of the Cleveland Bay as a foundation for breeding is one that could never be replaced; and the handsome and elegant Coach Horse or carriage horse is one we could ill do without. This recognition of the leading agricultural societies means much, therefore; it is, as it were, a new point of departure for the two breeds—from which they will attain to fresh excellencies and wider popularity.

CHAPTER IV.

THE ARABIAN HORSE.

THE following enthusiastic account of the Arabian Horse is from the pen of Captain W. A. Kerr, V.C. :—

"If centuries of unsullied descent, a masterful prepotency—a gift that long and stainless purity of breed alone can bestow—mien and bearing haught and high, rounded symmetry of form, the ability to travel far and fast, courage, and resolution to struggle and endure, highly-developed intelligence, a generous disposition, a constitution of iron, bone of hardest texture, sinews of steel and flinty hoof—if these go to make up equine perfection, then the true high-caste horses of Nejd, and those shapely steeds, of equal birth, bred in the plain between the two rivers that drink of the waters of the Tigris and Euphrates, are assuredly the noblest of their race. Were proof needed of the Arabian's far back and jealously-guarded pedigree, it will be found in his fixity of type, in the characteristic spring of the tail from the crupper. A Seglawia Jedrán, a Managhi, or any *aseel* Arab is distinct from other breeds, and could be produced from no other stock in the known world. All the pride of all his race in himself reflected lives! In him, 'strength and beauty have come together!' So pure and distinct is he of race, so great his power of heredity, that no matter how violent the contrast may be, how radical the cross-out, the mint-mark of the desert remains distinctly visible through several generations.

"If the Arab lacks the grandeur of physique of such noteworthy specimens of the so-called thoroughbred as Wild

Dayrell, West Australian, Bend Or, Minting, and of a few other past and present worthy representatives of the three great Eclipse, King Herod, and Matchem lines, we must bear in grateful recollection that all of these full, broad streams had their source in the Darley Arabian, the Byerly Turk, and the Godolphin Barb—said to have been of pure Managhi descent. It would be passing strange (seeing the conditions under which they have been respectively reared) to find the Arab of the same stature as our English racehorse. From the days, perhaps, when the sons of Noah, descending from Mount Ararat, settled in the plain of Shinar, or from those of Nimrod, that mighty hunter, the great grandson of the first navigator, 'the bold man of great strength of hand, who stayed and tyrannised in Babylonia,' down to the present day, the Arab has been bred under circumstances well calculated to arrest his growth, and to inure him to long days of continuous toil, semi-starvation, and thirst. For a few months, possibly, he may enjoy the pastures of the Maharaina, of Esdraelon, or some watered plain in 'Araby the blest;' but for many more he has to subsist on scant feed, such as a Basuto pony alone could thrive upon. 'Never let an animal lose its sucking flesh,' is an axiom with our breeders, who are careful to keep their colts and fillies in growing condition. The Arab foal, on the contrary, endures great privations, has to follow its dam on many a forced march, and must pick up a living as it can, aided only by a little camel's milk when this can be spared. Delightful as is Mesopotamia and the crisp clear air of the desert in the spring, during the protracted summer it is a foundry furnace, the almost perpendicular rays of the sun shooting down upon the brain and spinal column as though concentrated in the focus of a burning-glass. The air is charged with particles of fine sand, scorching as from the blast of an oven; the parched ground radiates fervent heat. Climatic extremes, free from humidity, however—for the winter, at night especially, is bitterly cold—and oft the scantiest of scant fare, are conditions calculated to produce a hardy, terse little horse, all wire and whipcord; but certainly are not likely to rear that massive animal so eagerly sought after in this country, and which, for downright hard work, away from his oats and old meadow hay, is so useless out of it—especially under a tropical sun.

"Till very lately the Arabian has been out of favour with our people. With us size covers a multitude of evils, and anything not over 15.2, no matter how big of bone and large of frame, is termed a 'little horse.' But now, thanks to H.R.H. the Prince of Wales, Mr. Wilfrid and Lady Anne Blunt, the Honourable Miss Dillon, and others afflicted with 'the Arab craze,' increased attention is being directed to this Eastern horse, whose descendants are now found to be distinguishing themselves pre-eminently on all the racecourses of the world, not excluding the trotting-tracks of America; and whose blood (no matter in what channel directed or with what plebeian puddle mingled) has ever brought improvement in some shape or other, but mainly in respect of quality, stamina, courage, nervous energy, ivory-like bone, tough hoof, and hereditary soundness. In the desert, roaring—the curse of our breeds of horses, from the thoroughbred to the farm slave—is unknown, and the absence of this unmusical propensity is of itself an undeniable recommendation. But Arabs frequently stand well above the normal 14.1 and 14.2 high. The grey Arab stallion, Smetanxa, the founder of the Russian breed of Orloff trotters, was a horse of commanding size and of unusual power. Naomi, now in the United States, measures 15.2$\frac{1}{2}$. The late Mr. Skene (when Consul-General at Aleppo) sent me two Anezeh mares, the one 15.1$\frac{1}{2}$, the other 15.2. It is on record that in 1729 an Arabian was in service in Norfolk, 'by size 15.3, and strength proportionate.' Later on, in 1762, Pettigrew's Grey Barb—the Barbary horses are descended from Arabians — height 15.1$\frac{1}{2}$, was also travelling in the same county. Aleppo, the Ormonde of his day on the turf of Bengal, stood 15.2. The Sakhur tribe, on the borders of Moab, have some well-grown mares.

"Thanks to the *Live Stock Journal*, and to its contributors, the important part played by the Arabians, the Barbs, and other horses of Eastern origin in founding our families of racers, trotters, and Hackneys, and even coach-horses, has been freely and forcibly illustrated. It has been conclusively shown that not only is the Arab the most potent factor in the composition of our racers, but also that—

> 'Oh he's such a one to bend the knee, and tuck his haunches in,
> And to throw the dirt in flats' eyes he never thinks a sin.'

We know that one of the most famous trotting families of America is of Eastern descent, and that an astute and enterprising breeder, Mr. R. Huntington, of Rochester, New York, is so impressed with the value of this hard, blue blood, that he has made some purchases from Mr. Vidal's stud, with a view to replenishing this strain, and of establishing a pure trotting tribe, to rival the record of Maud S., of the Western gelding Gay, and of Axtell, in the States.

"The ancestry of this trotting phenomenon supplies abundant proof that Mr. Huntington is working on promising ground, and that to this colt's back desert blood is due the foundation of his excellence. That he has the trotting instinct intensified is substantiated by the following tabulated pedigree:—

AXTELL, b h, 3 years, 2:12.				
WILLIAM L	Geo. Wilkes: 2.22	Hambletonian	Abdallah	
			by imp. Bellfounder	
		Dolly Spanker	Henry Clay	
	Lady Banker, dam of Guy Wilkes 2:15¼	Mam. Patchen, bro. to Lady Thorne 2:18¼	Mambrino Chief	
			by Gano	
		Lady Dunn, dam of Joe Bunker 2:19¼	American Star	
			a Clay mare	
LOU	Mambrino Boy 2:26½	Mam. Patchen, bro. to Lady Thorne 2:18¼	Mambrino Chief	
			by Gano	
		Roving Nellie	Strader's Clay	
			by Berthune	
	Bird Mitchell	Mambrino Royal	Mam. Pilot 2.27½	
			by Lieut. Bassinger	
		by Comet Morgan	Sherman Morgan	
			by Buckshot	

"Morgan was an Anglo-Arabian, and the dam of Dolly Spanker, an in-bred Morgan mare; American Star owes his paternity to an Arab-sired Canadian pacer, and the name of the thoroughbred Stockholm crops up in his pedigree. Gano was by American Eclipse, who boasted the Arab strain, and both Sherman Morgan and Buckshot were doubly in-bred to Morgan. Henry Clay, a famous trotting stallion, was sired by Andrew Jackson (also a famous trotter), a grandson of the imported Barb Bashaw. So Axtell, sold some time ago for the sensational sum of 105,000 dollars, is anchored on a solid foundation of courage and endurance.

"Mr. Huntington's efforts will be watched with attention, and I, for one, heartily wish him all the success his imagination paints and his pluck deserves. A glance at the pedigree of many of the best trotters in America, prior to the Civil War, will show that the latent Arab trotting instinct was not slow in developing itself in the New World. The Russian Orloffs claim Eastern parentage. The Cleveland Bay, now so deservedly popular on the other side of the Atlantic, must have a dash of the Anezeh in him. The Romans had Arab races at Ebor, and the arrival of the Leeds and Darley Arabians caused such a furore that the nobility and landed gentry of the three Ridings vied with each other for the possession of this potent element. For years after this Yorkshire horses were practically invincible, till that famous Southerner, Whisker—said to have been as near perfection as need be—travelled north, and inflicted signal defeat on the stables of the Tykes. By the way, I must mention that Catherina (Whisker's famous daughter) ran no less than 171 races, and died at the age of thirty-one. What racehorse of modern days could stand such a 'bucketing' as this?

"Very few quite first-class Arabs reach this country, hence the unreasoning prejudice against them. The breed is said to have deteriorated within the last thirty years, but, seeing that it has successfully withstood several centuries of close inbreeding, this cannot possibly be the case. The Sultan has, it is reported, placed his veto on the export; but my impression is that, as in time gone by, I, with a well-lined purse and *carte blanche* as to price, could pick up a few specimens of the highest caste horses, and probably a mare or two, such as would silence all carpers, and convince the most sceptical that there are still to be found amongst the desert-born superlative illustrations of that stock which boasts an ancestry of great deeds and mighty traditions. Horses that outrival all the rest, 'the pearl of the casket,' are no more to be bought for a mess of pottage in Arabia than here; and this is especially true both in Nejd and Mesopotamia, where so much store is set by blue blood. As for the Sultan's dog-in-the-manger iradee, that writ won't run, especially when it is sought to be imposed on these nomad rievers, who care nought for the Sultan or the Khedive. The dollar (backed by influence) is as omnipotent in the horsehair tents of the Bedaween as ever it can be in Lombard Street. 'Gold wins where angels might despair.'

"The Arabian has somewhat suffered in public estimation by ill-advised attempts to measure his speed against that of our modern racehorses. I once entertained the fallacy, but a trial at Newmarket between a plater belonging to the late Mr. Caledon Alexander, and the famous Mysore and Bombay 'crack' Copenhagen, convinced me of my mistake. On the racecourse the Arab is at a manifest disadvantage, being built on lines which are quite different from those of our greyhound-like thoroughbred. He is bred in the desert, exists for one purpose, and that is to carry his freebooting owner on his long and rapidly-executed razzias; whereas the thoroughbred, since the days of King James I., has been the product of careful selection for racing purposes solely; and, in these latter days, every consideration has been sacrificed to the development of speed alone; all the science and sound principles of breeding, which our ancestors were the first to establish, being very generally disregarded. Had we started *ab initio* from pure Arabian blood on both sides, sire and dam

being alike noted for bottom and speed, and from different tribes, importing fresh blood from time to time for an out-cross; and had we so bred for the turf exclusively, there is no doubt that the Anglo-Arabian would now be in every respect as high under the standard as the average of our racehorses, of at least equal speed, and their superior in courage, soundness, and general utility. That this exotic breed can, in course of time, under altered conditions of climate, food and treatment, and does increase in height without sacrifice of power and just symmetry, is equally true. Miss Dillon had a two-year-old 15 hands, on short legs, and its dam was barely 14 hands, its sire but 14.2½. Her famous jumper, Raschida, was 15.1½, and at Crabbet Park were two fillies, bred there, of the same height. In the course of three generations English-bred Arabs will lack nothing in respect of height.

"But it is as a hunter and war-horse, or both combined, that the Arab is at his very best. In old Deccan days of 'saddle, spur and spear,' what stirring camp-fire tales each hard-riding pig-sticker had to tell of the superb little nag that had carried him so gallantly over such breakneck ground, 'as though the speed of thought were in his limbs!' What a picture is the trained hog-hunter at the jungle side as the discordant yells of the beaters, floating down wind, proclaim the find, and announce the joyful tidings that the bristling banditti are afoot! Note the eager expression of his clean-cut patrician head and wide-thrown nostril; the bold, resolute eye; his noble bearing, as he intently scans the echoing hill-side! How his desert blood mounts like quicksilver, every vein charged to bursting, lacing his thin, high-bred skin! He stands motionless but for the quiver of suppressed ecstatic excitement, and an occasional spasmodic shifting of that truest telegraph, the delicately-pointed ear. 'Ready, ay, ready,' is depicted in every attitude; then, as the chase draws nigh, and the sounder pelts past him like a hailstorm, he spins round on his hind legs, and is after that great surly tusker, the last of the ruck, like a gyrfalcon in pursuit of its prey. Now comes a burst, trying to the full the speed, stamina, sure-footedness, cleverness, and jumping power of the horse, as of the nerve and strength of the horseman riding for the spear.

> Hard on his track o'er the open' and facing
> *The devil's own country*, the pick of the chase,
> Mute as a dream, his pursuers are racing,
> Silence, you know, 's the criterion of pace.
> (WHYTE-MELVILLE, *slightly altered.*)

" He who hesitates is lost ; for the mighty boar has a high turn of speed, is in rare galloping trim, and takes the rock-scarped nullah, the prickly-pear fence with its festering spines, and whatever comes in his way in his quick, resolute stride. What cares he for yawning fissured regur soil, for boulder-strewn ground, or for sheet rock as slippery as glass ? His thick hide is impervious to thorn of stunted babul or khair, to lance-pointed aloe, or sword-like wild pine. Though showing honourable scars—a wild hog will rip up a horse with as much ease and as little ceremony as a huntsman paunches a hare— the Arab, accustomed to the sport, will course the boar, sticking to him, following every turn and wrench, and going open-mouthed at him. Where the cloven-footed robber goes, there he follows, fearing no fall, refusing nothing ; his heart as much in the contest as if he were the rider and not the ridden. Weight appears to make little difference to these small horses, their short stride and quick recovery enabling them to carry men seemingly much beyond their scope.

" During the last few seasons Arabs have been distinguishing themselves with hounds. No day appears too long for them, no country too big, and they make light of the proverbial three days a fortnight. One mare, fourteen years old, that had never seen an English fence in her life, negotiated her way over a bank country at the first time of asking, in a manner worthy of an accomplished huntress. Naomi has the credit of having, with thirteen stone on her back, over and over again cut down her field in Suffolk. Miss Dillon's mare, Raschida, has carried off eight jumping prizes in public contests, and that lady's three stallions, El Emir, Eldorado and

Maidan, all possess the 'lepping' power. The last-named won an important point-to-point steeplechase in India. Mr. Blunt, I believe, hunts some of his Arabs with the Crawley and Horsham Foxhounds.

"Our campaign in Egypt must have convinced the military authorities that British-bred horses cannot withstand the exigencies of an Eastern campaign. Without venturing on the domain of foreign politics, it is permissible to hazard the opinion that when England is again called upon to draw the sword in a great war, the scene of the land operations will be not very far distant from the head of the Persian Gulf. Very much of the unsoundness which affects the usefulness and durability of our horses is undoubtedly due to the transmissible or hereditary defects both in sire and dam. The softness is attributable to the blood, which wants rehardening; and to the coddling-forcing system in vogue with our breeders of thoroughbreds. This artificial method of rearing blood-stock has caused the superintendent of Government breeding operations in India to lean very strongly to roadster stallions; which in my opinion he erroneously describes as of a fixed type. We cannot deal too guardedly with our horse supply. It is a monstrous delusion to imagine that we can take up a position and await attack. No successful campaign can be carried out in these days, unless every arm be as mobile as possible. Great activity, rapidity of movement, and boldness in seizing the initiative are essential. Mobility implies an ample supply, both in the field and in reserve, of serviceable horses of a quality and fibre suited to the country in which the army is operating. As in the hunting field, so for military purposes, we must have 'blood on the top.' It is claimed for them that, in India, they get more bone than the Arabs—possibly they do; but is not quality sacrificed for quantity; hard, fine texture for less dense and durable material? How do the produce of the two compare as to endurance and constitution—which of the two work best under the tropical sun, and how about roaring?

"India's chief requirement is an unlimited supply of good brood mares of a special type. Out there dogs run to nose, and horses to leg. Such a class of mare as would meet her wants used to be bred in Cape Colony some years ago. These mares were full of Arab and Barb blood. The original breed of South Africa came from Spain. In Queen Elizabeth's time the Duke of Newcastle considered the Spanish horse superior to the Barb of Morocco. 'The Barbes,' wrote that authority, 'were the gentlemen of the horse kind, and Spanish horses the princes.' The grandees had evidently made good use of the blood introduced by the Moorish Sultans of Granada. During the eighty odd years since we ousted the Dutchman from the Cape of Good Hope, many thoroughbreds from England and Arabs from India have been sent thither. The climate being all that can be desired, breeders produced from this material a long, low, powerful, 'blocky,' sound, active horse, up to great weight, and of as good constitution as temper. The enormous wear and tear of horseflesh during the Indian mutinies depleted the colony of these excellent general-utility horses; but what it has produced in the past it can surely supply again. Mr. Melk's compact and shapely 'Kaapsche schemmels'—a pair of which frequently sell out there for £300—testify to the Cape's horse-breeding capacity. If the Indian Government insists upon Hackney or roadster stallions—the expression is of the widest latitude—then let us humour the whim, and breed them here and in South Africa for the State and the native rajahs, now taking an interest in the enterprise; but by all means let the blood be hardened by reversion to the Arab sire. Let the mares be of the big bony sort, not too high under the standard, but let us be careful that they have at least as much of the thoroughbred element in them as has Bourdass' Denmark 177, or Mr. Burdett-Coutts' Tom King. Some roadster mares are to be found with grand shoulders, high set on beautifully-turned quarters, fine Arab-like heads, and comparatively free from those tell-tale long hairs in their pasterns. A bit of the much-admired round wheel-like action might well be dispensed with.

Anglo-Arabian, Kiialed.

"A better stallion by far would be the Arab, on the big, roomy, thoroughbred, weight-carrier mare. By big is not meant one such as that giant of his generation, Arrandale, standing 17 hands, with bone in proportion, and, for one of his weight, certainly a wonderful light and easy goer. A well spread young mare, 15.3 to 16 hands, is quite tall enough; but there must be power enough to carry fifteen stone in the Shires. We want to breed Anglo-Arabs such as Colonel Gore's Moodkee, first prize in the hunter-stallion class at the Royal at Dublin, or Mr. Lofft's Gidran (bred in Hungary), a horse that has got some excellent stock. In the South African campaign, Colonel Gore, then commanding the Inniskilling Dragoons, rode a full brother to Moodkee, and the Arab blood told, for this charger was never sick or sorry, lasting out four picked horses his master had also brought with him from Ireland. Stallions so bred, and not brought up like fatlings, would nick admirably with colonial mares such as are advocated above. The Bernais, who are extensive breeders, prefer the Anglo-Arabian stallion to the thoroughbred, finding the foals by the former much stronger and easier to rear on their coarse fodder.

"The 'sealed pattern' according to which we should endeavour to breed stallions for the Indian Government, and for our own home general purpose use, will be found in the engraving of a portrait of the Anglo-Arabian so admirably depicted by Mr. P. Palfrey. In it will be traced a strong resemblance to the famous Sultan, the worthy representative of the Woodpecker branch of the King Herod line, a horse said to have been cast in an enlarged mould of the Darley Arabian, and in many of his characteristics, a reflex of Flying Childers. The line carried from the point of the elbow, along the belly to the stifles, is, it will be noticed, very nearly straight, as is the case with Ormonde, and is common to very nearly all blood Arabians. The deep back rib, which always takes away from the appearance of a deep brisket, is indicative of that stamina and constitution possessed by the 'air drinker of the desert.' For such a class of horse the Italian Government would be an eager competitor, and from Buenos Ayres large orders might be confidently anticipated. Khaled is good enough to hold his own against all comers on the flat or between the flags, to carry a first flight fourteen-stone man in the Shires, would mount the Prince of Wales to perfection at the head of the Tenth, or Lady Clara Vere de Vere in the Row, and could not be passed over by the most critical judge for one of the Queen's Premiums.

"This account is embellished with a portrait of 'Speed of Thought,' a pure Keheilan-Seglawi Arabian, bred by the Gomassa tribe of the Anezah, the property of the writer. 'Speed of Thought' was a dark, rich chestnut without white, save a star. His near eye had been knocked out by the point of a lance in a razzia. Height 14.3, girth 72 inches, measured 8¼ inches below the knee, and stood on perfectly-shaped feet, tough as the nether millstone. He was possessed of superlative quality from head to heel, of great muscular development; sinews clean and hard as pin-wire, and stood fair and square on the best of limbs and joints. High couraged, as proved when he beat the famous horse Long Trump by a short head after a desperate race; full of what the Americans term 'vim'; a strong, vigorous galloper; his bold, free, and jaunty walk, quite up to five miles an hour, being the theme of general admiration. Across country, though somewhat headstrong, he was as clever as a cat, and would face anything, no matter how big, how yawning, and on parade bore himself bravely, as became his ancestry. Great depth through the heart, strong shoulders, a muscular neck with marked breadth in front of the withers and immediately behind the ears, denoting lung, staying, and weight-carrying power. The nearest approach I have seen to this *aseel* son of the desert was Count de la Grange's 'Consul,' the winner of the French Derby, but this undoubtedly clever and shapely thoroughbred lacked his fine fibre and finish. A better constitutioned, sounder, or gamer horse never looked through a bridle, and

ARAB STALLION, SPEED OF THOUGHT.

'Hagar's desert Ishmael's plains' never reared his superior. A true *Sh'rubah Er'rech* (wind drinker) was he, such a one as the brilliant Khaled, 'the sword of God,' the long-armed Tarik, or the chivalrous Saladin might have bestridden in some wild charge for Islam. Among his four-and-twenty victories, all achieved in the best of company, may be instanced: The Poona Derby, the Dealers' Plate (Bombay), H.H. Agah Khan's Cup, H.H. Alli Shah's Cup, the Gold Cup (Bombay), the 100 Gold Mohur Sweepstakes, the Drawing-room Stakes, the Welter, the Forbes Stakes, and the Winners' Handicap; his racing career closing with the easy defeat of the crack Madras horse, Risaldar, in a three-mile match. The artist, as is his wont, has done justice to my prime favourite."

CHAPTER V.

THE AMERICAN TROTTING HORSE.

This handbook deals chiefly with English breeds of light horses, but in view of the origin of the breed and the interest that is taken concerning its performances, a chapter on the American trotting horse will not be out of place. In his valuable little work on " Horse Breeding," Mr. J. H. Sanders says :—" Our American horses are largely permeated with the blood of the English thoroughbred. Many of the best stallions and mares in England have been imported to this country, and their influence is seen on every hand. It enters largely into the ground work of all our trotting strains, and it is doubtful if a single great road horse or trotter has been produced in this country that did not possess a large share of this royal blood as a foundation upon which the trotting superstructure has been built." It is clear that in the eastern districts of England trotting matches were quite common during the last century. After quoting the performances of the celebrated Hackney mare Phenomena, in 1800, Mr. Leslie E. Macleod, in an exhaustive paper on the " National Horse of America " (printed in the United States Report of the Department of Agriculture, 1887), says—" The conclusion is forced upon us that the English had the material from which to build and evolve a great breed of trotters."

The fact seems to be that the Americans commenced the sport of trotting at the point at which it was broken off in England, and, using our materials and their own, they have brought it, and the breed engaged in it, to great perfection.

Thoroughbred Stallion, Mambrino (Foaled 1768).
(Sire of Messenger, from whom many of the best American Trotters are descended.)

A State law of New York, passed in 1802, forbidding all horse racing and trotting, was amended in 1821, and allowed the training, pacing, trotting, and running of horses upon certain regulated courses in Queen's County on Long Island. There were somewhat similar enactments in other States, and though racing was prohibited trotting was permitted, so that the sport received encouragement while racing was proscribed. In this way American tastes were guided in the direction of trotting, a circumstance which has had no small influence in its subsequent development.

"The foreign horse that played the most important part in originating the American trotting breed, and that figures in the ancestry of our greatest sires and performers, was imported Messenger." In these words Mr. Leslie Macleod, in the paper to which we have already referred, confesses how large a part this great sire has performed in the creation of the trotter. Messenger is registered in the General Stud Book as having been got by Mambrino from a mare by Turf, from Regulus mare by Starling out of Snap's dam by Fox. He was foaled in 1780, and was grey in colour, like his sire, his height being 15.3. It was in 1788 that he was exported to Philadelphia, United States. His sire, Mambrino, was bred by Mr. John Atkinson, of Scholes, near Leeds, in 1768, and was sold in 1771 to Lord Grosvenor. He was got by Engineer (a son of Sampson) out of a mare by Old Cade, and it is stated that Mambrino was also sent to America, and "became the progenitor of the finest coach and trotting horses ever produced in any country, while, before quitting England, he begot some coach-horses that were never equalled." Sampson, the grandsire of Mambrino, was a black horse of great size and power. Lawrence observes that he was the strongest horse that ever raced, and was entitled to pre-eminence if viewed as a hackney or hunter. He was fifteen hands and a-half in height, and it is alleged by Lawrence that at twenty, and perhaps fifteen stone, he would have beaten over the course both Flying Childers and Eclipse. Sampson was foaled in 1745, and is entered in the Stud Book as got by Blaze (a son of Childers) from Hip Mare by Spark. A doubt was raised by Lawrence as to the accuracy of his pedigree on the dam's side, it being asserted that the mare appeared to be about three parts bred. Sampson was exceptionally large in size, and resembled a coach-horse rather than a thoroughbred.

As has been stated, Messenger was imported to Philadelphia in 1788. He was kept in Pennsylvania and New Jersey for the first six years of his life in America, and was also on Long Island, in Dutchess, Winchester and Orange Counties, New York and New Jersey, until his death, near Oyster Bay, Long Island, in 1808. "It was," says Mr. J. H. Wallace, "the crowning glory of his twenty years' service in this country that he left a race of driving-horses of unapproachable excellence, and, as he inherited this quality from his sire, so he imparted it to his sons, and they in turn to theirs, until we have to-day from this stock the fleetest and stoutest trotters in the world."

The most famous son of Messenger was Mambrino (named after his English-bred grandsire). He was a bay, foaled in 1806, being out of a daughter of imported Sour Crout. He never raced, and was so little valued that history loses trace of him for part of his career. He died in Dutchess County in or about 1831, and was a large, coarse, leggy horse, with well-defined trotting action. But he was a successful sire, and his son Abdallah, foaled in 1823, was sire of Rysdyk's Hambletonian, from whom some of the best trotters are descended. His dam was a daughter of the imported Bellfounder, of Norfolk blood. Rysdyk's Hambletonian is described as a bay horse of excellent structure, but very plain, the large head and Roman face especially rendering him objectionable to the eye of the lover of form. His most noted sons comprise Alexander's Abdallah, Aberdeen, Dictator, Electioneer, George Wilkes, Happy Medium, Harold, Messenger, Sentinel and Volunteer. Mambrino Chief was a grandson of Mambrino, his sire having been Mambrino Paymaster. He is the head of a family that ranks next to that of Hambletonian. Mambrino Chief was foaled in Dutchess County, New York, in 1844, and was from a mare of unknown pedigree. He was a fast trotter himself, and got some noted sons.

The Clay family of trotters was founded by Andrew Jackson, a trotter of high class in his day. He was a son of Young Bashaw, son of Grand Bashaw, a Barb imported from Tripoli in 1820. Young Bashaw's dam was by the racehorse First Consul, and his grand-dam was by Messenger. The dam of Andrew Jackson was a mare of unknown blood that, it is said, both trotted and paced. Andrew Jackson was foaled in 1827 at Salem, New York, and died in 1843.

Other blood-influences in the early record of the trotter were the imported stallion Diomed (winner of the first English Derby), and the imported Norfolk trotter Bellfounder (Jary's). The latter was foaled in 1816, and was by Stevens' Bellfounder out of Velocity by Haphazard. Velocity trotted on the Norwich road, in 1806, sixteen miles in one hour, and Bellfounder trotted in 1821, at five years old, two miles in six minutes. The Messenger and the Bellfounder blood was united in producing Hambletonian.

The Morgans are a very old trotting family; they are descended in the paternal line from a horse called Justin Morgan, bred in Vermont in 1793. The Pilots trace from the old black pacer Pilot, who was of French Canadian ancestry. He was the sire of Pilot, jun., sire of Maud S. and Jan-Eye-See, two noted performers.

The chief families of trotters, therefore, are the Hambletonians, the Mambrino Chiefs, the Clays, the Morgans, the Bashaws, and the Pilots.

The first recorded trotting performance in America was that of Yankee, at Harlem, New York, July 6th, 1806. The time of the mile was 2:59, but the track was not a full mile. At Philadelphia, August, 1810, a "Boston Horse" trotted the mile in harness in 2:48½. In 1832 Burster trotted in 2:32.

In 1834 Edwin Forrest lowered the technical record to 2:31½, on the Centerville Course, Long Island. Mr. W. H. Brewer gives the following table of increasing speed. We add to his list the more recent performances:—

1818	Boston Blue	3:0	1874	Goldsmith Maid	2:15½
1821	Top Gallant	2:43	1874	Goldsmith Maid	2:14¾
1824	Top Gallant	2:40	1874	Goldsmith Maid	2:14
1824	The Treadwell Mare	2:34	1878	Rarus	2:13¼
1830	Burster	2:32	1879	St. Julien	2:12¾
1834	Edwin Forrest	2:31½	1880	St. Julien	2:11¾
1845	Lady Suffolk............	2:29½	1880	St. Julien	2:11¼
1849	Pelham	2:28	1880	Maud S.	2:10¾
1853	Highland Maid	2:27	1881	Maud S.	2:10½
1856	Flora Temple	2:24½	1881	Maud S.	2:10¼
1859	Flora Temple	2:23½	1884	Jan-Eye-See	2:10
1859	Flora Temple	2:22	1884	Maud S.	2:09¾
1859	Flora Temple	2:21½	1884	Maud S.	2:09¼
1859	Flora Temple	2:19¾	1885	Maud S.	2:08¾
1865	Dexter	2:18¼	1891	Sunol	2:08¼
1867	Dexter	2:17¼	1892	Nancy Hanks	2:07¼
1871	Goldsmith Maid	2:17	1892	Nancy Hanks	2:05¼
1872	Goldsmith Maid	2:16¾	1892	Nancy Hanks	2:04

Other good records are those of Axtell 2:12, Allerton 2:09½, and Directum 2:05¼.

These figures show how the time required to trot a mile has gradually been reduced. A good deal might be said in reference to the changes in the formation of the tracks and to the use of pneumatic-tyred sulkies, but space will not permit of detail on these points.

The following notes on the breeding and management of trotters were written for us some time ago by Mr. R. C. Auld, Chicago:—

"Secretary Tracy voiced the popular sentiment when he declared 'that to get trotters, you must breed to trotters; and to attain the highest possible rate of speed at the trotting gait you must continually blend those strains that possess the greatest stamina and nerve force with those that possess the highest form of trotting speed. I am a believer in a thoroughbred foundation in the trotter, nevertheless, provided that it came from the best sources. I do not, however, like it so close up as do Mr. Robert Bonner and Senator Stanford. Safety in breeding lies in matching like with like. In other words, if you desire the highest type of trotter or runner, mate only with the best and highest type. As you cannot gather grapes from thistles, neither can you expect to breed world-breakers in point of speed at the trot from the Percheron or Shire horse, nor uniformly from a type, rich though it be in Oriental blood, whose instinct is to run and not to trot. As regards gameness and stamina, it does not appear to me that the highest type of trotter can borrow anything from the thoroughbred. The resolute manner in which he trots heat after heat, day in and day out, frequently after exhausting scores before getting away from the wire, leaves nothing to be said on the question of gameness.'

"John Splan, one of the leading trainers, writes: 'When we get brood mares with five or six crosses in their pedigrees, that have been tried through fire by actual battles on the Turf, bred to stallions with the same characteristics, we will, I think, have a family of racehorses that will not have to look to a pacing family for speed, or to the thoroughbred for staying qualities.'

"The veteran driver Turner declares: 'The horse that tries to win is the one we want, whether the breeding be gilt-edged or otherwise.'

"Pedigree can only give opportunity a better chance. It is a true saying that success depends on being ready when one's opportunity comes. Pedigree is the best means of making the trotter ready.

"There are fifty mares in the table of great brood mares whose breeding is unknown, and one hundred and fifty the breeding of whose dams is unknown. These facts illustrate what pedigree may eventually do in breeding the trotter. It will be some time before horses will have become so bred that they have all become classified in their various ranks so that none will be a trotter but of trotting breeding. The time might come, however, when the admission to the Register on standard claims will become occasional—if the American tendency to run everything into certain stud-book channels continues. This we much doubt. Every light-harness horse the American breeds, the first question about him is, 'Can he trot?' So that there will continually be additions to trotting ranks from other sources of light horses—as the Hackneys, Coachers and Saddlers—by occasional experiments of trotting sires on such females, or *vice versâ*.

"Speaking of the 2:30 class, a high veterinary authority holds that training for trotting predisposes to disease, and that there is more probability of finding some capital blemish in a trotter of speed than there is in others. This is but natural, however; the same holds good, doubtless, in regard to the thoroughbred. Breeding ought to be, in both, a safeguard against this predisposition. In the breeder's consideration of this question it will be at once seen what a use 'pedigree' is to him; it teaches him what strong lines to draw to, which weak ones to discard. It is here at once seen that the 'deeper in' he gets the safer he must be.

"It is said that the thoroughbred blood has always been a resource to draw from for staying powers. The late Senator Stanford put into training a thoroughbred filly, well-named Experiment, on the trotting turf. Considering the distinct anatomical conformity of the two goers it would seem as if a thoroughbred could not stand the pounding he must endure on the trotting track. On this subject we may quote old reliable John Lawrence, 1809, who says: 'It is a remarkable fact that there has existed no instance of a thoroughbred horse being a capital trotter. They soon become leg weary, and their legs and feet are too delicate for the rude hammering of the speedy trot.'

"'The advocates of the various theories of breeding,' remarked the editor of 'Wallace's,' 'are each finding their grain of comfort in the unparalleled records of 1891. The trotting purists claim the magnificent performance of the phenomenal two-year-old Arion, 2:10¾, and world's race records of Nancy Hanks, 2:09, and Direct, 2:06 (pacer), as upholding their theory. Those who believe that a thoroughbred should not be found closer on the pedigree than the second dam, point to the queen of the trotting turf, Sunol, and her magic 2:08¼. Those who believe in the sustaining power of the thoroughbred through the first dam, dwell lovingly on the champion stallion record of Palo Alto, 2:08¾, a half thoroughbred. 'Honours are easy.'

"Yet there was no chance work in the breeding of these. 'Every one was bred for a trotter.' Merit, therefore, does not seem confined to any particular line of good breeding. 'Good breeding in blood lines from good individuals, and from producing progenitors is the secret of success.' But it does seem that as far as thoroughbred blood is concerned it is best when not too close up. But the better the record its possessor has he is always deeper in trotting blood, which argues the eventuality of the thoroughbred trotter.

"While on the subject of breeding, allusion may be made to the recognition of the pacer in the Register. The pacing gait is that seen in the camel—the lateral propellers move together. In the trotter it is the diagonals that move together. Among some there is also an outcry over the pacer; he is— would be—ridiculed. But he is there to stay, and the best should be made of the fact. The fact is, further, that the pace and the trot are interchangeable. It is what such pacing stallions have done, not only in the way of siring 2:30 trotting speed, but also in the prepotency in speed production of their sons and daughters that makes it evident that it is impossible to disregard the influence of the pacing element in the trotter. 'Horses that sire pacers also sire trotters. Pacing stallions get lots of trotters. The sons and daughters of these pacing stallions keep on imparting speed at the trot to their descendants.'

"The conformation of the American trotter is noticeably peculiar to a foreigner. He is not drawn out so finely or whalebone-like as the thoroughbred. He does not stand— extend himself—over so much ground. He is of more stocky, compact build; has more sloping pasterns, a shorter and wider neck, so that through it plenty of air can be pumped to supply the deep capacious chest. His face is fine and intelligent—so that if a person had to choose a horse by one point he might select a trotter so. He is wide between the eyes. He is rather low in the withers. He has powerful hind-quarters, specially powerful hocks; these are noticeable as it is therefrom, as initial points, that his great bursts of speed emanate. His motion is peculiar. His hind propellers give one the impression of being thrown inside the line of the front propellers. He must have a straightforward gait, not swinging, which may mean loss of time by curving outward. He has not only to do the distance in a certain time, but he must do it in a certain way. There must be no going off the feet or breaking; all 'hitching,' 'skipping,' 'running behind,' is not trotting. A true trotting horse is possessed of nerve, judgment, self-control and determination. The trotter's steady, regular pounding of the turf—like the sound of the obsolete paddle-wheel in water—when it comes on the ear so synchronously and rhythmically, almost blended into one continuous sound, is the sweetest music on earth to the trotting expert.

"A description of Sunol may not be uninteresting. It has been said of her that it is a 'deuced lucky thing that she has a record.' She would never impress the beholder as being one of the fastest trotters in the world. Looking at the little bay mare, with her apparently heavy head and tucked-up stomach, one could almost persuade oneself that she had 'levanted with another's baggage,' and was travelling under false pretences. Her conformation curiously reminds one of the shape of the greyhound. She has the same deep chest; her stomach is drawn up. At least a portion of her head, particularly her ears, suggest the greyhound, while the sloping hips and slender steel-like legs add to the suggestiveness of the picture.

"It is a grand thing for the trotting queens and kings of these times that they have such a friend as Mr. Robert Bonner. In his stables, at West Fifty-fifth Street, New York, they

reach a haven of rest after their arduous trials, which is an honoured humane retirement. Mr. Bonner never races these world-beaters for money, seldom for exhibition.

"Sunol's daily diary may be noted. When she rises in the morning, she is given two quarts of oats: in the course of an hour she receives a drink of water. Then her groom brushes her lightly all over, puts on her walking boots and a light blanket, and takes her out for a walk of half-an-hour's duration. Returning to her stall, she is rubbed down, her boots changed. She is then hitched up and turned over to her driver. On returning from this exercise, she is rubbed with cloths until perfectly dry; a blanket is thrown over her, her boots removed, and her legs wound with soft flannel bandages, and she is walked slowly about to cool off. Again returning to the stall, she is rubbed once more with cloths and brushed until her coat shines, fresh bandages encase her legs, and a fresh blanket her body. Then she is about ready for some attention to her 'inner' wants—a hot bran-mash, followed by hay. Thus her morning passes. In the afternoon she is ready to receive visitors or take a spin in the park, driven by Mr. Bonner. Sometimes she may be hitched up with Maud S., but we imagine each would look better apart. Sunol weighed 1,070 lbs., which was a gain of 135 lbs. during the season.

"It will be seen how much grooming enters into the routine of Sunol's life. Grooming has, indeed, always entered largely into good horse hygiene. As in the days of Columella, it still seems to be considered that 'it was more beneficial to horses to be well and thoroughly groomed, than to be largely fed,' and that, without proper dressing, the horse could not attain that perfection of which he was capable.

"We may conclude by a reference to the recent changes in the management of the American Trotting Register, &c. Formerly the Register was conducted by Mr. J. H. Wallace and his company. The Register was begun by Mr. Wallace, and the first volume was published in 1868. A few years ago the breeders determined to acquire control of the Register for themselves; this would have been a simple matter, if there had been only one organisation of breeders; but there were two, hence a deal of competition arose as to which should gain control of it. The two Associations were, the National, with an adhesion of 400 track members, and the American, with an adhesion of 700. To make a long story short, Mr. Wallace, as president of the old Register Company, finally gracefully surrendered to the American, as the stronger element. He was paid 150,000 dols. for the copyright of the Register, Year Book, his Monthly, and plant. The American Association, whose offices were formerly in Detroit, moved to Chicago; there conventions will meet; from there they now issue the various publications connected with the Register, including the Monthly, which has certainly gained largely by the change."

The following rules, for registration of standard trotters, came into force in April, 1893:—1. "Any stallion that has a record of 2:30 or better, provided two of his get have records of 2:20 or better, and provided his sire or dam is already a standard animal. 2. Any mare or gelding that has a record of 2:25 or better. 3. Any mare that has a record of 2:30, provided her sire is standard and her dam is by a standard horse. 4. Any stallion that is the sire of four animals with records of 2:30 or better, or the sire of three with records of 2:25 or better, or two with records of 2:20 or better. 5. Any mare that has produced an animal with a record of 2:25 or two with records of 2:30 or better. 6. The progeny of a standard horse, when out of a standard mare. 7. Any mare whose sire is standard, and whose first and second dams are by standard horses."

CHAPTER VI.

THE HUNTER.

THAT the thoroughbred is the foundation of nearly all our half-bred stock, is simply a truism. The blood horse makes the best sire for our hunters, hacks, chargers, troop horses, and for those harness horses which are bred from Hackneys, Cleveland Bays, or Yorkshire Coach Horses; though in the case of the last-named type, the thoroughbred may claim an additional amount of credit, since he has been called in to add quality to the Cleveland Bay. With these breeds, however, we are not now immediately concerned, and so may confine our attention for the present to the hunter.

For hunting purposes, no horse equals the thoroughbred, provided only that the rider does not too heavily tax the mechanism of the weighing machine. So long as the hand does not pass 11 stone 7lbs. on the dial—which means that the extra weight of saddle and hunting clothes will not bring the total to more than 13 stone—no man need despair of riding a blood horse in the hunting field; and when he has once ridden him, he will never want to go back again to any half-bred horse. The longing to keep to the *pur sang* will prompt him to mortify the flesh, if needs be, in order to keep down his weight; for when a man has once experienced the easy, elastic gallop of the thoroughbred, he will not readily adapt himself to the more laboured action of the half-bred.

It is sometimes said that the thoroughbred cannot jump so well as his relative of commoner lineage, and that he takes longer to school; but both these statements may be dismissed with the remark that they are inaccurate. So long as a thoroughbred horse can be schooled to jump the Liverpool course, the hunting man may comfort himself that, if his own heart be in the right place, he never need be pounded in any county in England. At the several Newmarket sales, and also at Doncaster, yearlings are sold at prices varying from 15 guineas to 30 guineas, and in the opinion of the writer it would be well worth the while of any light weight to purchase some of these, to turn them out, and "forget all about them" for a couple of years. About three pounds of oats per day—and they need not be of the very finest quality—would go far towards building up their frames and fitting them for the duties of the hunting field; a slow racehorse is a very fast hunter.

The time is probably very far distant, when men will breed thoroughbreds to hunt; but there does not seem to be any reason why—if any one chose to try the experiment—thoroughbreds up to weight, should not be reared. One sometimes sees both thoroughbred horses and mares with great bone, and for hunting purposes, to carry weight, it would be useless to think of breeding from anything which had not substance. There are generally some hunter sires going about having quite sufficient bone, but there might be a difficulty about getting the proper type of mare; though, judging from the prices realised for unfashionably bred ones, the difficulty need not be an insurmountable one, and after one or two mares had been bought, the fillies they might throw could, of course, be utilised at the stud, if they proved suitable. Considering the pleasure to be derived from riding a blood hunter, and remembering that one able to carry 14 or 15 stone would always realise a large price, it is perhaps rather astonishing that so very few people have tried the experiment of breeding the blood hunter. It may be objected that to do so would be embarking in a profitless speculation, inasmuch as more money would be given for a yearling to race than for hunting. True; but this only holds good in the case of those with more or less fashionable lineage, and everyone who has attended the bloodstock sales must have heard Mr. Tattersall trying to obtain a bid of 20 or 25 guineas for a yearling by some unknown sire, and out of a dam who has not yet made a name for herself. When these youngsters come to more mature years, and turn out too slow for racing, they find their way into cabs, or are sold at low prices for other work; whereas had they been treated in a manner calculated to fit them for becoming hunters, they would, at four years' old, have been worth four times the money that would be given for them as Turf failures.

The embryo race-horse is trained, galloped, and tried, and these processes sometimes result in rendering him unsound, when of course he is practically valueless; but under the more gentle *régime* of the paddock and the hunting stable he would never be asked to gallop as a two-year-old; his work at three years would be of a light description only, consequently his frame would have time to get well set before it was taxed by work. It is not, of course, pretended that every young thoroughbred would remain sound, or make a valuable hunter; but it stands to reason that if they are not set to severe exertion in their two-year-old days they have a greater chance afforded them of growing into sound horses; so that the proportion of failures owing to breaking down must be smaller than in the racing stable. When the time comes for the future hunter to be schooled over a natural country a fresh set of risks begin. He may break his back at some little ditch, or otherwise injure himself; or he may be unable to stand the strain jumping puts upon his legs; but these things have to be chanced with every horse, thoroughbred or not.

So long as we deal with thoroughbred animals alone, whether they be horses, cattle, sheep, or dogs, we may breed to type; but directly we come to crossing one breed with another we are landed in a sea of uncertainty. We may have a thoroughbred horse on the one side, and a good looking hunter mare on the other: they are mated, and the produce may be worth £100 at three years old, or it may be fit for nothing better than cab work. It may be 16 hands high, or may never grow beyond 15 hands. The breeding of half-bred stock, therefore, may be truly regarded as a lottery; though at the same time there are certain rules and fixed principles which should not be lost sight of.

Beginning with the sire, it has been laid down as a general rule that he should be thoroughbred, or practically so. A slight stain in the pedigree should not, however, disqualify a stallion that is otherwise suitable as a hunter sire. The writer has seen one or two good hunters got by a trotting sire out of well bred mares; but one would think hunters so bred are rare. Experience has shown, too, that in the majority of cases medium-sized sires are more successful than very tall ones. Soundness is, of course, a *sine qua non*, and so are good limbs; while it is as well to ascertain whether the horse it is proposed to use has got his mares in foal, as failure to do this puts the breeder to much expense and loss. If the sire has acquitted himself respectably over a country it will be a recommendation; but it is not so necessary that he shall have won races as that he shall have shown himself a good fencer, for jumping capability often runs in families like temper, pace, and other attributes; while for a hunter it will be no harm if the sire be somewhat of the "cobby" order, so long as his shoulders are well placed, and his back and loins muscular.

The reader is probably aware that a number of persons are greatly in favour of Arabs as hunters, and of using Arabs as hunting sires. So long as a horse is of proper make and shape, can carry the necessary weight, and jump properly, it does not matter to the *user* how he is bred; the man who breeds only to mount himself can ride a jumping bull if he likes, as did Jemmy Hirst. But when breeding for sale is the object, the breeder must try to produce the animal that will sell best; and it may be questioned whether a half-Arab is the sort of horse after which buyers will run. We read of Arabs of 14.3 and 15 hands carrying 13 stone and upwards over all sorts of

country; but, as a rule, as mentioned elsewhere, size means power, and it is a succession of big fences that beats the little horse. It may be that the 15-hand Arab is equal to the English horse of three or four inches higher, but the majority of hunting men would be slow to assent to the claim; consequently the breeder who can offer a buyer nothing but small horses must be prepared to pay for his fancy in the shape of a reduced price. It is not contended that the cross between an Arab and a hunting mare is necessarily small, but usually they give one the idea of being nothing more than light weight horses.

The breeder cannot too soon realise the fact that the choice of a suitable sire is only one step towards breeding a hunter. It is of paramount importance that the dam should be equally good in her way. Yet many breeders, small farmers especially, when they are not themselves great horsemen, persist in breeding from weedy, undersized or worn-out mares. It is not until a mare is past work that some of them think of sending her to the horse. The very natural result of this is that nine out of ten of the produce are fit for nothing better than to put in a butcher's cart, indeed, sometimes they have not pace enough for that; and then the breeder exclaims that horse breeding is a delusion, and that after all the expense and trouble have been undertaken he has had to sell a four-year-old for about £20 or less. The wonder would have been if the animals had brought any bigger prices. Even when the greatest judgment is exercised—when both parents are just what one would think they ought to be, and when everything is done to bring on the young stock—there must be a certain number of failures; but when judgment and prudence are cast to the winds, who can wonder if disastrous results follow?

In the first place it is no more worth any one's while to set out with the idea of breeding light-weight hunters than it is for them to lay themselves out for breeding a stamp of horse that shall sell for £25 or £30. In the ordinary course of things the breeder, even if in the long run he be successful, will find himself with as many misfits as he wants; his aim should, therefore, be to produce a horse of the highest class, and this, in the hunting department, may be described as a horse up to 15 stone with hounds, and with as much quality as possible. As often as a good sample of this sort of horse can be bred, so often will a remunerative price be forthcoming either for the raw material or when the horse shall have become a finished performer over a country. In a certain proportion of cases the breeder will be so far disappointed that he will find he has a light weight horse instead of a weight carrier; consequently there is no reason why he should try to breed the latter unless he confine his attention to thoroughbreds.

The careful breeder will do well never to breed from a mare of whose history he is ignorant. She may have been put to a Hackney, Cleveland Bay, cart-horse, Arab, or half-bred sire; and it is a well established fact that a mare, like the female of other animals, will frequently throw back to the male with which she was, at some anterior time, mated. If, therefore, a mare happens, unknown to her present owner, to have been put to a cart-horse, she may, when mated with an eligible thoroughbred, drop a colt with the quarters of a blood-horse and the head and neck of a Dobbin; and though this may not affect the horse's performances, it will affect its appearance, and what is more to the point, the price. The first thing, then, is to ascertain the history of the mare from which it is proposed to breed, and if this be impossible, it will be best to leave her alone, or, at any rate, not to pin one's hopes on anything she may breed.

What kind of mare is most likely to help her owner to breed a weight-carrying hunter is a question which it is practically impossible to answer. A mare, which to outward appearance is just what a hunter brood-mare should be, is necessarily made up of several strains; and her produce may take after some of her ancestors just as they may favour some horse with which she may have been mated, as mentioned above. If cart

Hunter Mare and Foal.

blood predominates, the offspring may be a heavy, shapeless thing, fit only for a van. If, however, the mare has been ridden, and if she rides lightly and gives one the idea of being well bred, the chances are in favour of her not throwing back to anything coarse on her own side. When there is a predominance of cart-blood, there is also present, as a rule, a heavy action and a general kind of clumsiness which can be detected by any one in the habit of riding well-bred horses. The texture of the coat, too, is sometimes another guide to the mare's fitness to become a brood-mare.

Strength she must have, in the form of both bone and muscle; and she should not be less than 15 hands 2½in. in height, nor should she exceed 16 hands 1in. In this, as in other matters, the mean is best; and perhaps 15 hands 3in. is about the best height for a brood mare. At the same time, although size generally means power, it does not follow that a tall horse is necessarily up to weight—a self-evident proposition, yet one which a good many breeders do not appear to have grasped. There was a good deal of sense in the remark of the old master of fox hounds who declared that the "height of a hound had nothing to do with his size." This is true of horses to a certain extent; but as we presently propose to show, a certain amount of height is as necessary in a horse as it is in a hound.

To return to the brood mare, however. Various experiments have been tried. With the hope of combining strength with fashion, thoroughbreds have been crossed with heavy cart mares, and the fillies so produced have been again put to the blood sire, and so on; but the result has scarcely been satisfactory, and after the second cross the progeny has come out in all sorts of shapes. Others have tried the clean-legged cart mare, with better results. In former days, when the breeding of hunters was at its best in Ireland, the dams of the hunters were almost invariably the clean-legged cart mares of the country, for five and thirty or forty years ago hairy heels were practically unknown in Ireland. Then by degrees the

Clydesdale and Shire horses were introduced, to take the place of the native mares, which had been sold to go abroad, and it may be stated with confidence that since that time Irish hunters have not been what they were.

What we require—in theory, at least—is an upstanding, big-boned, roomy mare, got by a thoroughbred horse, and with as few mixtures in her pedigree as possible. Such a mare is hard to find, and when found will not always throw the sort of foal we want. A few years ago, "G. S. L.," an acknowledged authority upon breeding matters, ventured the opinion that every hunter brood mare should have some pony blood in her, on at least one side of her head, and there certainly seems to be a great deal in this theory, when one calls to mind the mares one has known which have been descended from New Forest, or what are now regarded as Exmoor, ponies.

At the best, however, the writer regards the most eligible stamp of weight-carrying hunter as, under existing conditions, a purely chance-bred animal, and no rules can be laid down which will give a breeder a reasonable chance of thinking that he can breed two in succession from the same parent. We are speaking now of horses equal to 15 stone at the most; but there is still more chance about breeding those elephantine animals which can be ridden by men who walk from 15 stone to 17 stone. These must give up all hope of quality, for the man who rides 17 or 18 stone to hounds must be thankful to be carried at all, and must be grateful for the assistance of an active cart-horse. On the other hand, however, one has seen some wonderful heavy-weight horses. When Lord Macclesfield used to hunt the South Oxfordshire country, he must have ridden 16 or 17 stone; yet what horses he had! The writer well remembers two of them in particular, a chestnut and a brown. Both had the quarters and middle piece of a dray-horse, but they could gallop at a great pace, and jump anything, and were by no means "carty." Then again, Mr. Merthyr Guest, Master of the Blackmore Vale Hounds, and Mr. Heywood Lonsdale, Master of the Shropshire, ride horses showing a wonderful amount of quality for the weight they have to carry; but all these horses we believe to be chance-bred ones.

We have only to go to horse shows to find out how rare are weight-carriers with quality. Out of about a couple of dozen entries, at least one quarter will be voted not up to the minimum weight; about the same number will be common, and only fit to follow harriers in a sticky, slow country; it will be possible to find faults more or less serious with some of the remainder, and when the judges come to make up their minds how the prizes shall be awarded, their choice will probably be limited to three or four. It will be observed, too, that as the limitations as to weight decrease, the classes grow very much larger. This has been less observable lately than it was a few years ago, for latterly, the tendency at horse shows has been to diminish the number of classes assigned to horses capable of carrying various weights. In connection with this question, the following extract from an Irish newspaper of 1886 may not be without interest:—" That our horse-breeding has changed within the last twenty-five years is evident by the great present scarcity of horses to carry over 14 stone, and the increase of those who can carry 12 stone. In last year's show (1885), although most substantial prizes were offered by the Royal Dublin Society, the proportion of classes were as follows:—For weight-carrying hunters up to 15 stone, five years' old, 64 entries; hunters from 13 stone 7 lbs. to 15 stone, five years' old, 97 entries; while for the class for hunters up to from 12 stone to 13 stone 7 lbs., five years old, there were 143 competitors. Now, comparing this with the show held in 1876, exactly ten years since, the entries in the then respective classes were, 15 stone class, 25; 13 stone 7 lbs. class, 46; 12 stone class, 30."

Reference to the catalogues of our English horse shows give very similar results. For example, when the Royal was held at Nottingham in 1888, there were a dozen heavy weight hunters and seventeen 12 stone horses; at Hull, in 1889, there were eleven heavy weights and 21 light weights. At Islington, in 1889, there were 21 horses in the class to carry a minimum of 15 stone; 28 in the middle weight class, and 25 in the light weight class.

The moral of all this is, that while hunters up to weight are difficult to rear, it is comparatively easy to breed light weight horses, and that these should exist in such large numbers proves that the breeding of weight carrying hunters of high class is chance work. As already stated, no breeder possessing common sense would lay himself out to breed light weight horses in preference to weight carriers, because experience has shown that the ranks of light weight hunters are very largely recruited from failures. A man who rides from 12 stone to 12 stone 7lbs can mount himself cheaply enough, since he may ride either a weed or an undersized horse. In every country in England are light men who get along and maintain a good place on something not exceeding 15 hands; and if one goes to Tattersall's, Aldridge's or to any other repository, it is possible to see plenty of competent light weight hunters knocked down at sums not exceeding £70, while a great many bring no more than £30 or £40, or more properly speaking, guineas, while some are obtainable at still lower prices.

In the matter of general rules, therefore, we can get no further than saying that the hunter brood-mare should possess both size, strength, and breeding; and when we have all these three requisites we must still be indebted largely to chance.

This brings us to the consideration of the question what is a saleable horse? To a certain extent we have discussed this question already. He must have size, breeding and quality, and, of course, jumping abilities of a high order. The "horse for Leicestershire" is, in short, the horse to bring the most money; and whatever the theory of individuals may be, in practice every one who can afford it buys a horse of this type, no matter in what country he may hunt. Go to Northum-

Hunter.

berland in the north, Sussex in the south, Lincolnshire and Essex in the east and Devon in the west, and you will find that men who have the money—or the credit—will mount themselves on well bred, upstanding horses able to gallop and jump. Of course on Exmoor and Dartmoor where there is no jumping, in Kent and Sussex where big woodlands are met with sufficiently often to cause checks, and in close, rugged countries wherein climbing and creeping are more the rule than galloping and jumping flippantly from field to field, a good deal of sport can be seen from the back of a horse which would be of no earthly use in a grass country in which are small coverts. But a few times in a season hounds run hard in the worst of countries, and then it is that the value of a good horse is seen. To put the matter shortly, no man rides a worse horse if he can afford a better. Jumping a country is merely a matter of local practice, and there is no reason whatever why a horse which can get over a grass country should not, after a little practice, be an equally brilliant performer over the wide Roothing ditches of Essex, the banks of Dorsetshire, the stone walls of Gloucestershire, or the formidable ramparts of Devon. In every country in England horses are seen which would not disgrace themselves on the grass anywhere.

These well bred, strong horses, then, are the ones the breeder wants to produce if he can, as they bring the most money. In the majority of years each individual will have to remain content with a lower standard and consequently a lower price, and what he fondly hoped would turn out a high class hunter may eventually have to be sold as a harness horse.

The writer has always advocated that at horse shows the brood-mares should be divided into two classes. When numbers of mares of all sorts and sizes are shown together, the prizes naturally go to those showing most quality. It would be well, therefore, to have one class for mares themselves up to 14 stone, at least, with hounds, and the other for mares up to not more than 14 stone. In this way the heavier mares, which may be the better calculated to breed weight carriers, would have more chances of gaining prizes than they have at present. It would also be a move in the right direction if greater rewards could be given in the classes for farmers' brood-mares. If a farmer with a decent mare could nearly make sure of picking up £25 or £30 a year at small shows there would be forthcoming proof positive that keeping a better mare paid best, not only because she could win prizes where a worn-out thing could not, but also because better young stock could be bred.

Hunter Sires.

In an article on "Hunter Sires," Sir Walter Gilbey, Bart., makes some suggestions which are calculated to render the breeding of hunters more reliable than it has hitherto been. He writes as follows:—

"It is strange to observe how satisfied many people continue to be with the present system of breeding general-purpose horses. This condition of mind affords a striking instance of the influence which deep-rooted prejudice, and a determination to adhere to fashion, can exert upon the intelligence of men. The idea has prevailed too long that weight-carriers—or useful riding and driving horses—can only be bred by using thoroughbred sires; and this belief has proved the chief obstacle in the way of every suggestion for a better system of breeding strong and sizeable animals.

"There is no doubt that the thoroughbred of to-day is, for racing and for reproducing speedy animals, better than he ever was before; yet the fact that there has been great change in rules and customs of the Turf has made him, of necessity, less suitable as a sire for getting horses for weight-carrying, for harness, or for military purposes. The altered system of racing and modified Turf arrangements have combined to produce

A Hunter Sire.

a class of racehorse which is quite distinct from the thoroughbreds of last century. We have now horses that come quickly to perfection, and as quickly pass from the Turf. At the present time there are not a dozen races of any importance which are run upon courses above two miles. The ordinary distance is from six furlongs to one and a-half miles, and the exceptions are a few welter races, which are still contested for by the better class of horse.

"The average weight which is carried, ranges from 8 stone to 9 stone. In very few cases is the higher of the two limits exceeded. This arrangement gives the good, bad and indifferent horse a chance of winning; and speed has been preferred to substance. Horses are tried at two years old, and if they appear to be slow they are at once cast, in order that the expense of further training may be saved.

"In a controversy (which has been going on for many years) it has been proved that no plan, for improvement of the stock of general-utility horses, can be long maintained by using, exclusively, sires reared for the Turf. Even many racing men now share this conviction; which has been to some extent brought about by the fact that several of our best four-mile steeplechase horses are not clean-bred; but have other blood running through their veins besides that of the thoroughbred. This is the case with the following steeplechase horses of high repute:—Roman Oak, New Oswestry, Zoedone, St. Galmier, Heather, and Marienbad. The first-named won many races in 1891, and the big race (£2,000), at Manchester, in 1892. In the list of Grand National winners, a large percentage will be found to have what is considered to be a stain in their pedigree. This is the case with Pathfinder (who won in 1875), with Zoedone (1883), Old Fox (1886), Gamecock (1887), Frigate (1889), Ilex (1890), and Come Away (1891). The Colonel (who won twice *i.e.*, 1869 and 1870) and The Lamb (who won in 1868 and 1871) both belonged to the same class.

"At the commencement of this century, competition in racing all over England, was carried out on quite different terms to those which are observed to-day. Then it was for four-year-old horses carrying 10 stone 4 lbs.; for five-year-olds with 11 stone 6 lbs.; and for a few aged horses with 12 stone, and it was decided in four-mile heats. Such races were a great inducement to breeders to endeavour to get horses of size and substance; and so long as these prizes and the Queen's Plates were given to horses carrying heavy weights, strong thoroughbred horses continued to be bred and kept upon the Turf.

"For many years past our stock of really sizeable riding horses, and of true-actioned, well-matched driving horses, has been notoriously deficient. To make up for this deficiency, large numbers are imported from abroad, and we are sending away vast sums of money which ought to go into the pockets of British farmers and breeders. The Government returns—for the six months ending June, 1892—show that there were 3,932 horses *more* imported than in the corresponding six months of the previous year; viz., 12,343 for the six months of 1892, as against 8,411 in the six months, January to June, 1891.

"It is curious to learn that we have been doing in the immediate past exactly what was done in this country between the years 1154 and 1702. In order to prove this, an interesting article, headed 'Antiquity and Progress of Horse Racing,' should be examined, which appears in one of the volumes of the *Sporting Magazine*, published in 1810. In this is briefly described the commencement of English horse racing. The article states that it was—

"'Only after the reign of Henry II. (1154) that gentlemen began, among other feats of sporting, to try the fleetness of their horses against one another. . . . Gentlemen went on breeding their horses so fine, for the sake of shape and speed only. Those animals which were only second, third, or fourth-rates in speed, were considered to be quite useless. This custom continued until the reign of Queen

Anne (1702), when a public-spirited gentleman (observing inconvenience arising from this exclusiveness) left thirteen plates, or purses, to be run for at such places as the Crown should appoint.* Hence they are called the King's or Queen's Plates, or guineas. They were given upon the condition that each horse, mare, or gelding, should carry 12 stone weight: the best of three heats over a four-mile course. By this method, a stronger and more useful breed were soon raised, and if the horse did not win the guineas, he was yet strong enough to make a good hunter. By these crossings—as the jockeys term it—we have horses of full blood, three-quarters blood, or half-bred, suitable to carry any burthens; by which means the English breed of horses is allowed to be the best, and is greatly esteemed by foreigners.'

" In the face of all that may now be seen and read, no one can deny the importance of this subject to the nation. Indeed, the question has for more than fifty years occupied at intervals the attention of Parliament, of the press, and of a large percentage of the horse-loving people of the country. In no age has the idea of perfection been placed higher than it is in this. It is admitted that the object of all should be to combine usefulness with beauty, and that there is, or should be, some visible standard of what is being aimed at. In the breeding of general-purpose horses, men should have before them some type or model of what they are seeking to obtain. Now, it will be found that there are in existence—to guide us—many old pictures of celebrated sires other than thoroughbred, who, in the early and middle part of this century, did good service by begetting progeny of the desired character, and who were freely used for breeding horses suitable for carrying heavy weights and coaching horses. And, what is more, there exist (in the old *Sporting Magazine* and other contemporary publications), the written records

---

* If this be a true statement it may be seen from what source the money came which originally supplied what—from Queen Anne being on the throne—were termed the Queen's Plates.

and particulars of this useful type of horse. At the period to which I refer sons of these weight-carrying sires were themselves continued as sires. Thus a course of line breeding was established, so that there came to be a breed of light horses quite distinct from the thoroughbred. Why men should not have persisted with this line-bred stock of horses it is difficult to understand. It can only have been through the craze for speed, without regard to other important attributes, such as size, bone, and usefulness generally.

"Among many other pictures of old hunter sires to which we might have made reference, there is one of 'Pantomime' in the *Sporting Magazine* of the year 1836. This is called 'a favourite hunter, the property of David Marjoribanks Robertson, Esq.' The dam is described as '*not* thoroughbred.' And in the article appended it is stated that 'the sire of Pantomime was Grimaldi, a race of hunters nearly extinct, and justly celebrated for their high courage, honesty, and stoutness.' What, it may be asked, became of that 'race of hunters?'

"With cattle and sheep it is fortunate that the practice has been different. We have had men whose enlightened minds led them to persevere throughout on the lines that their forefathers adopted, and the success which has thus been achieved proves line-breeding to be right. Bakewell, in the last century—the great pioneer in improving nearly all descriptions of stock—bred from animals which were on both sides of exactly the same character. He had no intake of fresh blood for upwards of twenty-five years, and the merits of his method are recognised by herd and flock-owners down to the present time.

"But in our attempts to breed hunters and heavy weight-carrying horses the idea has prevailed that with these no distinct type can be fixed in the same way, because no relationship should exist between the sire and dam. This arose from fear that the progeny would be too near akin, and that if mares of a good line-bred sizeable sort were put to a sire bred in the same way and of the same character as the dam, it would be doing wrong. In real fact, there is no danger in breeding by the nearest affinities, provided they have developed no unhealthiness, and provided the animals to be mated are both possessed, in a superior degree, of the qualities which are sought to be established.

"It is impossible to imagine why we should have postponed so long the breeding of weight-carriers on both sides from animals of a type that would reproduce itself. For we have acknowledged breeds in the thoroughbred, the Hackney, and the Shetland. In heavier breeds we have the Shire, Clydesdale, Suffolk, and Cleveland, all of which varieties reproduce themselves without recourse to any outside alliances, and there seems to be no reason why a breed —to be hereafter called the 'hunter sires' or 'weight-carrier sires,' or by any other name—should not be established in the same way.

"The haphazard system of continually using thoroughbred sires, and this to mares which already have a large proportion of thoroughbred blood, has not proved successful. Those who have expended large sums of money in trying to breed in this way weight-carrying hunters and sizeable horses have acknowledged their failure.

"To realise this truth, it is only necessary to read what Lord Cathcart has written in the nineteenth volume of the Royal Agricultural Society's Journal. The article occupies fifty-five pages, and gives, with much information, the opinion of numerous practical breeders.

"If my opinion were to be asked, and I had to advise briefly what should be done I would say, 'breed from a stallion, other than a thoroughbred one, which has a strain of hunter blood in his pedigree,' or select a thoroughbred stallion that possesses the shape and make of a hunter, and is capable of carrying a 14 stone man to hounds. If he be mated with a hunter-mare of known descent—one that has carried not less than 14 stone to hounds, has won hunter or point-to-point races, or that has won premiums at the Spring Shows of the Hunters' Improvement Society—then the progeny of such mating will be a commencement of establishing a heavy-weight line of hunters. There is no animal better for coach or carriage purposes or for work requiring powers of endurance, such as doing long journeys by road, than a horse of the hunter class. The result of such an experiment would be sizeable animals, which if not suitable for one purpose would be for another. Here the remark of a writer (Mr. Chas. W. Tindall) is much to the point when he says: 'We can make a hunter a harness horse, but all the Acts of Parliament cannot make a harness horse a hunter, and it is a fact beyond dispute that horses of the hunter type are more in demand than any other for general-purpose work.' "

Young Half-bred Stock.

It is well that those who undertake the breeding of half-bred stock should thoroughly understand that it is a risky business, and that a great deal of knowledge of one kind and another is required before it can be made to pay. It is also necessary to understand that where no more than two or three foals are bred annually, a loss must inevitably result if the breeder has to pay for labour, or rent of premises. If a man has the convenience for breeding, he may do fairly well in a small way with luck and good management. Attention is drawn to these matters, because they are not without their bearing on the case when the time comes to ask what is to be done with the young horse?

Clearly it will not do to pay to have him broken and made into a hunter by any one else, or all the profit (if any) will be swallowed up.

All young horses should be handled from the day of their birth, it saves a world of time and trouble afterwards—in fact, if a foal be constantly handled, and be early fitted with a head collar, and then with a colt bit, or better still, the ring bit, such as is generally seen in the mouths of yearlings at sales, and if a little rug be put over him at times, and if the attendant occasionally bear a little on him, breaking to harness or saddle will generally be attended with no difficulty whatever.

At this stage of the proceedings some very important considerations come in. The breeder may be an excellent man for breeding, *i.e.*, he may be a good judge of mares and stallions; he may be thoroughly up to the work of treating his mare properly while she is in foal, and he may know all about handling and taking care of young stock; but he may be no horseman, and may be by nature unfitted to undertake the task of breaking or making a hunter. In such cases, it is submitted, the breeder would do well to sell his youngster to the first person who will pay him a sum which represents a fair working profit. It may be urged that by-and-by this horse may be sold for £200, £300, or £400; so he may, but let not this consideration trouble the mind of the breeder. By selling young he has the minimum of risk, and possesses that bird in the hand which is proverbially said to be worth two in the bush. We have somehow come to regard breeding and breaking and making as one man's work, whereas it is really the task of two men, though of course some have the gift of doubling the parts; but when a horse is old enough to be sold as a hunter, the price paid for him depends upon something more than good looks: his manners and performances must be good if a large sum is to be given for him, unless, of course, he is good looking enough to be bought for show purposes.

Unless, therefore, the breeder or one of his family be a sufficiently good horseman to undertake the breaking and making of a hunter, the breeder is best out of his colt as soon as he can dispose of him at a profit, as if he does not carry himself well, move in good form, and jump freely, cleverly, and temperately, he will never pay the breeder for keeping till he is four or five years old. It may, however, be necessary to adopt a middle course, that is to say, the breeder may, through lack of offers, be unable to dispose of his youngster as a yearling or a two-year old, yet he may find a buyer before the colt is old enough to be ridden as a hunter. In this event some progress will have to be made in his education.

At two years old he may be taught to jump, on the plan recommended by Whyte Melville, that is to say, he may have to jump a low rail and little ditch to get to his feed of corn, which he should have once or twice a day; at two years old he may be ridden by a light weight, and at this early stage he should be taught to stand quietly while being mounted; to walk well, without breaking into that uncomfortable jog at which some horses will persist in travelling, and not to start with a jump. Directly a horse is found to be deficient in any one of these points his value falls; but a very moderate horseman should be equal to guaranteeing that his steed does not fail in these elementary particulars. During the preliminary lessons the colt will be held by an assistant while the trainer mounts; but as the colt grows accustomed to being mounted, the assistant will hold him less; but he should never be suffered to move till the rider has both feet in the stirrups, has gathered up his reins, and has given the signal to start.

The more a horse knows of jumping the better, and if he has been accustomed to jump his little rail and ditch for his food, he will have done something to develop the muscles which come into play when taking a fence, and he will have learned something of the art of balancing himself when taking off and landing. People are not all agreed as to the best method to be adopted to teach a horse to jump a "natural country," that is to say, hedges and ditches, water, &c.; but the writer is of the number of those who have implicit faith in the efficacy of a leading rein. He believes in it for two reasons. In the first place, the horse can have his initiatory lessons before he is old enough to carry a weight upon his back without danger to his legs; and secondly, he will get accustomed to banks, ditches, and other obstacles, without being incommoded by a rider who may possibly pull at his head a a critical moment, and so give him his first lesson in "stickiness"—a great fault in a hunter. At the outset, the leading rein—fastened to a caveson—is to be preferred to driving the horse over fences with long double reins.

It is important that only small places be selected until the colt will trot up and jump them readily. The trainer will at first be accompanied by an assistant, who may lightly hold the horse while the trainer gets over the first little ditch or hedge; and here the trainer may be advised not to turn round and stare the pupil full in the face when asking him to jump. He should stand sideways to the obstacle, so that he can see all that the colt is doing without frightening him. The same rule holds good when taking a horse into a horse-box for the first time. The person in charge should never turn round and look the horse in the face, but should precede the horse by a foot or two, and keep his back towards him. The assistant should be provided with a whip, but it is to be used sparingly, and not at all if possible—a slight cracking of it will nearly always suffice—and, above all, the trainer must not be in a hurry. If he set about his work properly, nineteen colts out of twenty will give no trouble whatever, especially if they have been in the habit of jumping the rail to get their food. Nor must the trainer be in too great a hurry to increase the size of the fences he selects; small places only should be picked out till the pupil jumps them without hanging at them.

It would be beyond the scope of these remarks—which are intended for beginners—to deal with the advanced education of a hunter. The person who undertakes to ride a young horse over a country should be a man of experience. If, however, the breeder be not a good horseman, and if he will educate his young stock on the lines here laid down, he may feel sure of being able to realise the best price for what is little more than raw material. The buyer will see that he has a good foundation to work upon, and will not have to spend a long time in unlearning several bad habits, which he would certainly have to do had some unfinished horseman undertaken the education of the colt.

There is no harm in initiating young horses into harness work; they may at first draw bushes or a log over a field. The writer once saw a two-year-old fitted with a breast strap, and helping to work a small mowing machine; he was not of much use, it is true, but, as the owner said, he was getting used both to drawing and the noise of the machine. On the same premises were a two-year-old and a three-year-old working a chaff-cutter. The breeder of these young horses, though by no means a great horseman, had the knack of effectually handling young horses, while he lost no opportunity of making them familiar with the sight and sound of trains, traction engines, clothes hung to dry on a hedge, and with the hundred other things which terrify horses. The man who breeds in a small way generally has plenty of time on his hands, and the gradual process, which may be almost likened to the Kindergarten system of educating children, is far preferable to the adoption of "systems" in which a colt has his head and tail tied together, or is thrown down by complicated tackle. Nevertheless, these things have their uses; and no harm will come if the breeder learns one of the systems, because in case one of the youngsters should turn out refractory, the mechanical appliances which enter largely into the systems come in handily in obtaining a mastery which is not easy to effect with a mere halter or caveson

After Alken.   "AWAY! AWAY."

CHAPTER VII

THE HACK.

SIR FRANCIS HEAD tells us that "to metamorphose a hack into a hunter is chiefly effected by the bridle;" and it may be equally true that the bridle has not a little to do with changing the hunter into a hack. At any rate the majority of hunting men prefer as a hack a horse built on the lines of a hunter. After being accustomed to plenty of length in front of the saddle, to good sloping shoulders, and the swinging action of the hunter, it is scarcely possible to bring one's self to ride a little roly-poly cob, short in front, and with high quick action.

For quiet riding on the roads one does not, of course, require a horse up to as much weight as the hunter ridden in the winter; but for most tastes the hack must be on hunter lines, even though we come down to the 14-hand polo pony. For park work we may desire more action than we should care about in the hunting field; but if anyone will visit the Row day after day during the London season he will see that high-stepping hacks are by no means common. The horse show hack is frequently an animal quite *sui generis*—he commonly bears about as much resemblance to the working hack as does the English nobleman in the opera " Marta " to members of the peerage as seen in ordinary life.

The cob appears to be, now-a-days, a nondescript sort of animal; and it is only here and there that one is seen which conforms to the requirements of a perfect hack. Even at the best of the horse shows nearly all the cobs are wide-chested, thick-shouldered, very short of quality, and with rough, uncomfortable action; yet so long as they can bring their knees sufficiently near their noses their owners think them fit to exhibit. Some years ago, in writing about cobs, an acknowledged judge of horses remarked that the requisites for this sort of hack were perfect quietness, good shoulders, a reasonable amount of length, manners and quality, and he continued, "If the cob can walk four miles an hour, trot with equal ease six or ten miles an hour, canter as slow as five miles an hour, then let the owner, if a rich man, cling to him, for he will not get such another in a hurry; if a poor man, let him not hesitate to open his mouth, and demand for his cob a sum which shall make an appreciable difference in his year's income."

It has often been said that a park hack should be highly trained to answer more readily than the hunter to aids and indications. No possible exception can be taken to this doctrine, provided only that the rider is likewise educated up to the higher development of equestrianism. A hack that suddenly stops, traverses, passages, or does something else on receiving an unintentional hint, would be nothing short of a nuisance to a rider who is not fairly well up in the details of *la haute école*. In common parlance, a hack may be too well broken; but above all things he should be docility itself. A man who is making a young horse may naturally expect his steed to sit up with him now and then; or in consideration of brilliant performances in the field a hunter which will not go quietly to covert may be tolerated; but for park work or for quiet road riding, lamblike placidity is a *sine quâ non*.

It is seldom or never that the hack—not being a covert hack —or some useful slave kept for the sole purpose of "supplying a want," and getting over the ground as quickly as possible, is required to travel fast; consequently pace is a superfluous quality in a horse required for the park. But a hack should be a good walker, and no pains should be spared to make him excel in this mode of progression. In teaching a horse to walk

he must be made to go into his bridle, as indeed he should in his other paces; the reins should be held one in each hand, so as to check the horse on either leg should he show the slightest inclination to break. He must not be taught, however, to bear on the bit for support, but should step away freely when he has learned his lesson, with the reins on his neck if necessary. If a horse walks too fast for others, he can always be kept back, but if he be a slow walker, his rider will rarely be able to travel in company, except at that jog-trot, the constant indulgence in which knocks twenty pounds or more off a hack's value, and upsets his rider's temper.

In like manner a hack should be schooled to carry himself in good form, and to trot at a pace not exceeding eight miles an hour, without wanting to increase his speed to the uttermost, and finally to break into a gallop. The slow canter should also be a studied pace.

In hacking, no less than in hunting, the choice of a proper bridle is a matter of the greatest importance. It was once more the fashion than it now is, to "provoke the caper that they seem to chide;" but this is a silly proceeding. Some men even now, affect what is known as a " hard and sharp " *i.e.*, a curb bit only, but this practice is not recommended to unfinished horsemen. No sound argument can be advanced in favour of a needlessly sharp bit, for an engine which frets a horse and gives him artificial showiness must be more or less cruel. For ordinary horses the ordinary double bridle will suffice for all purposes, and if from any reason—want of schooling or bad breaking, for instance—the horse should carry his head too high, the French, or standing martingale, may be employed, without any of the risks that would attend its adoption when using it over a country.

It has already been stated that a hack should be quiet, and quietness includes an absolute indifference to sounds, noises, and sights which may not be encountered at every turn. Scarcely a season passes without some accident occurring in the Row from a horse taking fright, owing to some other horse or horses passing it at a canter, or from a lady's habit flapping in its face. It is no difficult task to make a horse familiar with these things, and no hack can be considered as fit for use till he is proof against the fear or surprise which anything unexpected may engender. Throughout the length and breadth of England traction engines, railways, gipsy vans, and bands of music are so common that no matter where a horse may be bred, he should possess a good knowledge of the world by the time his preparatory education is completed. We may here warn both breakers and horsemen against a very common fault. Very many people on the approach of a train, a band, or on passing within reasonable distance of a shooting party, are at infinite pains to inform their horses that something is about to happen which may frighten them. That is to say, they pull up to a walk, begin to say "Whoa," take tight hold of the reins, and so far as acts can effect the purpose, give their steeds every encouragement to begin to prance and fidget before anything happens, and to make a halt when the sound is heard or the sight encountered. This, it is respectfully submitted, is wholly wrong and quite opposed to the common sense of horse-breaking. The effect of giving a horse a signal in advance is well illustrated in the musical ride at the Royal Military Tournament at the Agricultural Hall. After the cantering ride is finished the horses are drawn up in two lines at the upper end of the arena, there to await the sound of the trumpet which is the signal for the charge. Before the last horse has taken his place the others show their eagerness to start; they know what is coming, and can hardly be restrained. So it is with the civilian's horse; once gather up your reins and begin the whoaing process, and you plainly tell your horse that something is about to happen. The common sense plan is not to upset him by grasping with the legs, and tightening the reins, and beginning a nervous conversation with him. Sit as you were sitting, leave the reins as they were, and hold your tongue. You will, of course, be ready to restrain your horse should he attempt to get away; but if he starts there is no need to take it for granted that he is going to run away. If you yourself are not in a fright the chances are that your horse will retain his composure.

People who train shooting ponies and chargers know quite well that the best way is to leave the reins as slack as possible; the horse just starts at the unaccustomed report, but after a short experience he becomes quite indifferent to the sudden noise. If the reins were nervously clutched, and he were given to understand that firing was about to take place, he would be unsettled. Try the same sort of experiment among your friends with a bottle of soda water. Let them see that you mean to open it; remove the wire, turn the cork very slowly, and extract it more slowly still. You will then see some of your audience making grimaces, contracting their eyebrows, and giving other evident signs of the agony of suspense; and when the cork does at last come out with a pop, more than half of the company will instinctively recoil and jump; yet they do not do so at dinner when servants open champagne or aërated waters. Then they do not know when the noise is coming, and it is over before they have time to start.

In treating of horses, however, it is assumed that they have been gradually accustomed to different sights and sounds during their early breaking. Lastly, the hack, like the hunter, should be taught to stand as still as a rock while the rider mounts, and the rider, in his turn, on becoming possessed of a well-broken hack, should be very careful not to spoil him by accustoming him to start almost before he is settled in the saddle. It is the custom to abuse colt breakers, but more than one-half of the faults of horses are caused, not by the breaker, but by the mismanagement of those into whose hands they subsequently fall.

The Harness Horse.

It must depend upon circumstances whether a person about to set up a horse and carriage, gets the horse first, and then looks out for a suitable carriage, or whether the process of selection be reversed. It occasionally happens that people who are not too well off are able to get either a horse or a vehicle on unquestionably advantageous terms — sometimes for nothing. When this is the case, it is unwise to decline the offer; but in either event the one should suit the other in point of build and size. Nothing looks worse than to see a great camel of a horse drawing a low and light vehicle, and nothing looks more inhuman than to set a small and light horse to draw a great heavy carriage.

Where, however, any reasonable outlay is of no moment, the first thing to be done is to decide upon the purpose for which the equipage is required. If you want to go to the station two miles off in ten minutes, you will not, of course, give a couple of hundred guineas for a high-stepping carriage horse and drive him in a brougham; nor, if you propose to drive in the park, are you likely to select a buck board waggon, and an American trotter with a 2.30 record.

For a full sized landau or similar carriage, horses of 16 hands, at least, will be required; but it is a matter of taste whether they shall be of the hunter type—that is to say, the result of a thoroughbred horse and a half-bred mare—or whether they shall be more nearly related to the Cleveland Bay and Yorkshire Coach Horse. For country work it may perhaps be difficult to beat the hunter-bred horse; he has plenty of pace and power, while as he is not likely to have very high action, his legs will last all the longer. This lack of action, however, tells against the ordinary half-bred horse for park and parade purposes, and those who desire something more showy will be more likely to suit themselves by buying —or hiring—horses of the Yorkshire Coach Horse stamp. It is, however, impossible to divide harness horses into distinct

breeds, for, like the hunter, they are frequently made up of many strains. The sire of the largest will be either a Cleveland, Yorkshire Coach Horse, or thoroughbred, and sometimes a Hackney; while the dam may be of the above breeds, or some kind of a half-bred, with more than ordinary action.

For brougham, victoria or phaeton work, the Hackney blood is a good deal in favour, and not without reason. Than the squarely built horses about 15.2 one sometimes sees, with good, but not extravagant action, nothing can be more admirable; but people who desire to preserve that grand action—the possession of which adds so vastly to the price of a horse—must remember that steppers are not calculated to perform long or fast work. If many miles have to be covered daily, the carriage stable must be proportionately strong, for unless a horse be a little above himself, he will soon lose his action; while if he be driven at any pace, the chances are that he will not be long in battering his legs to pieces.

It is apparently a matter of conscience with the majority of coachmen to drive high-stepping horses with sharp bits and tight bearing reins; but the horse owner would do well to at once set his face against the abuse of these, at times, necessary engines. It was said just now that a horse will not continue to be a stepper unless he be above himself, and being above himself implies somewhat easy work, and generous—not too generous—keep. It is obvious, therefore, that a horse whose life is cast in such pleasant lines will be full of fire, and having regard to the fact that he will spend much of his time in the crowded streets, he would not often be a fit subject for a snaffle, or to be driven at the cheek. Without any suspicion of cruelty, therefore, we may safely adopt somewhat stronger bitting. The question is, what bit is best for the purpose? Many people like the Liverpool pattern, which is by no means destitute of advantages; but the horse's lip sometimes gets pinched by the shifting mouthpiece, especially in double harness, where there is always more or less lateral pressure on the bit. There are, however, other patterns, which are free from this defect, but unless a horse be some hot and heavy-headed brute that can hardly be restrained, the abominably high ports which disfigure so many harness bits should not be adopted. To provoke high action by such illegitimate means is sheer cruelty.

Then we come to the bearing rein, a piece of harness as useful in its way as a kicking strap. The well-fed and easy worked horse need not necessarily carry his head in the right position, and it would not do for him to extend himself too much in the streets of London. The bearing rein, properly used—not abused, mind—is, in the case of a headstrong horse, of material assistance to the coachman, and prevents horses in double harness from getting their bits or bridles fast against one another or on the pole. But there is not the slightest necessity to pull up a horse's head to the extent we unfortunately too often see, and directly the swivel to which the bearing rein is fastened is found to be a couple of inches or so above the mouthpiece of the driving bit, it may be taken for granted that the coachman knows but little about the science of bitting. To speak the truth, the average coachman is lamentably ignorant of bits and bitting.

The dogcart horse does not require much notice. He may be anything that will suit the cart in point of size, and the work required of him, so far as his pace is concerned. He should, however, have good forelegs and shoulders, inasmuch as if he falls, the occupants of the cart are likely to be hurt. A buggy horse should have both action and pace, and so should the cobs which are driven in the now fashionable varieties of two-wheeled cars and carts.

Some of the remarks we made in connection with the training of the hack, apply with equal force to the harness-horse. Nothing is more annoying than to have a horse which will not stand while people get into a vehicle, or which as soon as he has the signal to go, first rears up and then starts with a bound. This habit, which is easily taught, is with difficulty eradicated, and in order that it may not be acquired, the

Hackney Mares, Princess and Brunette.

coachman should never be hurried or flurried when mounting to his seat. The passengers should get in quietly, and the horse or horses should be started gently, without anything approaching to fuss. If haste has to be made to catch a train, it will be time enough to put on the pace when you have emerged from the carriage drive. A very few hurried starts, and a horse will be made fidgety.

CHAPTER VIII.

PONIES.

The difficulties which beset a writer on the subject of ponies are rather formidable. In the first place, it is no easy matter to state definitely what a pony is, in a fashion that is likely to prove acceptable to breeders in the various districts of horse-breeding England. In the opinion of many an experienced man, anything below 15 hands is a pony, but this dictum, it need scarcely be added, is simply the rankest heresy in the judgment of the majority of pony raisers. Then, again, it may be parenthetically observed, the remarkable number of divisions and sub-divisions which ponydom includes, affords a puzzle to the uninitiated, which it must honestly be confessed requires a great deal of explaining, even by past masters in the art of rearing this class of stock. Nevertheless, for the purposes of this chapter—which in no sense is intended to be a scientific article on our little horses—it will be simply sufficient to allude to the principal varieties of pony in a less analytical fashion than would be imperative in a more ambitious work. The uses of ponies, the principal designations by which the leading breeds are recognised, and their breeding, may be lightly touched upon.

To commence with, therefore, the writer will endeavour to point out to his reader, and convince him, of the immense value of these equine bantams which to the discredit of our countrymen have for so long a time been permitted to languish in the obscurity of unconsidered trifles. There is hardly any branch of usefulness for which a pony is not adapted—short, of course, of carrying heavy weights or drawing ponderous loads. It would be sheer lunacy to pit a pony against a large horse in an equal competition, but if tested pound per pound it is pretty certain that the little one would hold his own and a bit more, assuming that the handicap were fairly carried out. One has only to consider for a minute or two what sort of work it is that many a costermonger's pony is put to day after day, to arrive at the conclusion that in a competition David *versus* Goliath, the giant would come off second best if the weights and dimensions of the two animals were carefully ascertained and their respective burdens adjusted accordingly. A pony, moreover, is supposed to live principally upon air, as will be shown in the part of this chapter that will deal with the question of breeding him; whilst, in countless instances the meagrest shelter is afforded him, and even this is of the roughest kind.

It may, therefore, be assumed that one of the many virtues possessed by ponies is their utility; a second, the smallness of their appetites; and, lastly, the peculiar propensity they possess for "roughing it." In seeking for spheres of usefulness which are particularly adapted for ponies, one has to go no further than the countless tradesmen's carts which throng the principal streets in every town throughout the length and breadth of the country. Between the shafts of these conveyances every variety of nondescript animal which can by courtesy be styled a horse may be met with within the space of half-an-hour's walk, but it is well nigh equally certain that the vehicles which are travelling best and look smartest are drawn by ponies. Animals like the three-cornered, mad, Argentine wretches that have been put into their carts, with the most unfortunate results, by some fatuous tradesmen, who simply could not get suited—at their price—with any other class of horse, are not at all the sort of animal to leave standing outside an area gate in a crowded thoroughfare, and, moreover, they are far more costly in the long run than a pony, if only on account of their larger appetites, and the extra price that has to be paid for harness and traps to fit them. The number of ponies, moreover, that are at work, in pairs, in the London streets has sensibly extended during late years, for van owners and the like, who must have their work transacted with reasonable despatch, and who cannot afford to have their animals blundering and falling about the streets, appear at last to have arrived at the opinion that a pair of 14 or $14\frac{1}{2}$ little ones—this is rather transgressing the pony standard, by the way—are far better adapted for this purpose than wastrels standing a hand and a-half higher at the shoulder.

Of course, too, when it comes to the question of working in coal mines, the little ones are quite in the front, but their suitability for such work need scarcely be discussed in a general notice of ponies, as it concerns a very small section of the community—mine owners, to wit. On the other hand, the thousands of mouths which are daily dependent for bread upon the labours of the costermongering fraternity would fare but badly if the latter had to rely solely upon the humble donkey as an animal of draught. At the exhibition of ponies and donkeys, *bona fide* the property of London costers, which took place at the People's Palace, in July, 1892, the owner of a pony informed the writer that he had tried donkeys, but found them quite unsuited for the particular work his animals were expected to perform, and he added with emphasis " There's one or two of my mates who's a lookin' for a good pony, but they can't find 'em." This forcible allusion to the dearth of useful ponies that exists in London, is one that might be borne in mind by owners of waste land, where they might be inexpensively reared, as when a demand exists it is only reasonable that some efforts might be made to supply the same.

Having thus alluded very briefly to a few—but only a few—of the rather prosaic, but perfectly legitimate uses to which ponies may be put, some reference may be made to the loftier walks in life to which their energies and capacities may be adapted. In the first instance, it is obvious that the splendid game of polo would become an impossibility, were no miniature steeds available and to hand for the accommodation of the players. Nor could children learn to ride, or old gentlemen be safely carried in the park or on the moors, if something up to weight and steady, and near to the ground, were not provided to do their will. Then there is the legion of timid individuals and invalids, who would shudder at the bare idea of finding themselves on two wheels behind a horse, but to whom the pony is quite a thing of joy and comfort, in which implicit reliance may be placed. Here, indeed, is to be found a market, which it is impossible to believe that an increased supply of the *proper* sort of pony—not the goose-rumped Russian specimens, or the cow-hocked productions of some districts of England—could possibly glut; and moreover, it may reasonably be anticipated that the popularity of ponies would extend, as a knowledge of their value became more widely recognised.

Finally, some allusion must be made to the highest class pony of all, and the most valuable, the trapper, with all round action and good looks to attract the attention of the uninitiated and horsey men alike. How these little animals are appreciated may best be gauged by the infallible test, the money one, for the way that heads go nodding when a top-sawyer comes up under the hammer, shows how difficult they are to procure, and how highly prized they are when come across. Any breeder who could turn out goers of about 13.2, and good-looking to boot, would never be likely to receive an unpleasant communication from his bankers regarding the state of his balance, and with reasonable luck and judgment would be a rich man in the course of a few years. Yet with the knowledge of this fact before them, British horse breeders have contented themselves with confining their attention to the bigger breeds, with the result that the appearance of a new pony at an Islington or Agricultural Show, is regarded as an event in the annals of the equine year.

The second advantage which may be assigned to the pony over heavier horses, is the smallness of his appetite, and this should prove no small attraction in the eyes of would-be equestrians, afflicted, alas! as so many of them are, by ill-lined purses. Whether the pony is naturally so small a feeder as he usually is, is, however, questionable, as there can be little room for doubting that the early treatment which the majority of equine bantams experience has provided them with a facility for thriving upon very little. All excess in eating and drinking is due to the inordinate encouragement of a natural appetite, and if this is rigorously checked it is marvellous in how short a time the craving will disappear. Ponies, usually, owing to the mere fact of their existence and their earliest surroundings, are well acquainted with the art of going upon short commons, and thriving on the same. This circumstance in unquestionably due to the belief—and it is a true one—that young stock that are well fed grow, and that those whose dietary is limited do not. Consequently, from his earliest infancy a pony finds himself compelled to try and get as fat as possible upon an amount of food per diem which animals more happily circumstanced could dispose of in half the time, or less. It is hard work, too, for many of the little colts and fillies to rub along during their most youthful days, as the majority of their breeders very sensibly endeavour to arrange so that their mares foal late, in order that the youngsters should derive just as much nourishment, first from their dams, and then from the grass, as will enable them to live and move and make as little growth as possible.

Doubtless, therefore, the natural appetite of a healthy pony becomes reduced by habit, with the result that, when he reaches his full stature, he flourishes upon, and is contented with, a far smaller allowance of food than he would otherwise require. This faculty for going upon short commons must surely be regarded as an additional recommendation, which should never be lost sight of by those who are hesitating between investing in a horse or pony for purposes of pleasure, and who may not be blessed with ample means.

Having been accustomed all their lives to rough it, the majority of ponies readily adapt themselves to circumstances which would be objectionable, and prejudicial to the health of other horses. Of course no animal can be expected to thrive or do well in an ill-ventilated, stuffy stable, but a pony can stand fresh air better than any other horse, and in this respect should prove attractive to a buyer. This hardiness cannot, however, be received as a proof that the little ones are not benefited by care and attention in their stables, for no living creature ever flourishes if neglected. No owner need expect that his animals will show good coats if left unattended to, and it is only natural that the spirits of a pony that gets good corn will be higher than his neighbour who gets none.

In the case of a pony breeder, it is one perpetual war against an increase of size on the part of his stock, and to ward against this he has to see that they are inbred, foaled late, kept on short commons and not pampered. In offering this opinion the writer most certainly has no intention of implying that the animals should be starved, or subject to any sort of cruelty. In the first place, the nature of any right-minded man would shrink from either proposing or adopting such advice; and secondly, it would be extremely bad policy were they to do anything of the kind. At the same time, all stuffing and coddling must be rigorously tabooed, or the ponies will soon become horses, and disappointment will ensue. In order to emphasise this opinion, which has been founded not only upon personal examination and experience, but also upon the ideas of many leading breeders who have been consulted upon the point, the writer may so far enter upon a digression as to invite his readers to consider for one moment whether the steady increase in stature which is so obvious in most, if not all, breeds of English horses, is not directly the result of high feeding and scrupulous attention? It may also be remarked that ponies are always commoner in sterile districts than are horses, and this circumstance is most unquestionably due to the fact that their food, and that of their ancestors, does not contain the nutritious elements that are to be found in rich pastures and luxurious lowlands. The little ones are unquestionably the children of the mountain and the uplands, and it therefore must assuredly become the bounden duty of a breeder to endeavour to imitate the teachings of Dame Nature, and not oppose her by stuffing his youngsters with food which no mountainous district could by any possibility supply.

In making the assertion that ponies are the "children of the mountain," the writer desires it to be clearly understood that he does not imply that they cannot be bred on, or may not be indigenous to, the lowlands of any country. At the same time no dwarf breed of animal will retain the smallness of its stature if raised for generations on a fertile soil and under a genial climate. A 16-hands Arab, or anything like it, would be a *lusus naturæ*, but it is a certainty that these desert-created steeds are steadily increasing their height at shoulder so far as the English-born animals are concerned.

It has been observed before in this chapter, that in-breeding has to be practised in order to maintain the desired smallness of a pony, and there is not much chance of the truth of this assertion being combated by practical breeders. The very history of the most famous varieties, be they Dartmoor or Exmoor, New Forest or any other, confirms the accuracy of the statement, and it usually occurs that a change of blood at once has the result of raising the height and increasing the substance of a strain, and continues to do so until by systematic "sibbing" the effect of the new blood has worn itself out in this respect. Considerable differences of opinion exist as to how these in-breeding operations should be conducted, but speaking generally, it is more advisable to breed father to daughter and mother to son, than to adopt the cross of brother and sister. Unquestionably most of the recognised varieties of pony—such as those raised on Dartmoor—are very much in-bred, and doubtless the juveniles would be of a far better class than they are if some rational method of management were adopted. It is no use to emasculate the most ill-favoured of the young horses, and then let the others serve the mares haphazard, or at all events if it is so, the less that is said about scientific breeding the better. Nature in a large number of cases is a grand mentor to be guided by, but unfortunately her beneficial laws are just as likely to be set at defiance by ponies as by men, and with equally disastrous results. Consequently a judicious control over the animals that are running at large is always desirable, if only to ensure the proper amount of in-breeding.

Having now ventured to obtrude upon his readers his opinions upon the value of ponies generally, and the best methods for procuring the desired reduction in size, the writer finds himself compelled to approach the consideration of that much-vexed question, "What is a pony?" If this enquiry could be answered by the simple reply, "Oh, anything under 15 hands, or 14.2, or 14," or whatever it might be, the difficulty in offering a solution of the riddle would be simply *nil;* but unfortunately there is no such an easy means of escape to a man with a pen in his hand. He, at all events, is compelled to offer some more definite solution of the problem; for it must not be imagined for a moment by the inexperienced that the correct definition of a pony would be, a horse that stands a certain number of inches at the shoulder. Of course all ponies must be small, so far as their height is concerned, but having agreed upon this point, one at once is assailed by difficulties of many and varied kinds. A cob, for instance, is a small horse, but he is not a pony, strictly speaking, and there is almost as much difference between a good specimen of the stocky weight carrier so beloved of elderly gentlemen possessed of a tendency towards *embonpoint,* and a bloody-looking natural mover, as there is between a Shire horse and a park hack. It becomes necessary, therefore, to endeavour to arrive at some clearer definition of what a pony really is than that of a horse that stands beneath a certain standard.

Unfortunately at the present time, the several varieties of pony are suffering from the want of proper care and attention which should have been received by them from English horse breeders, and it would be impossible to attempt a critical analysis of the points and characteristics which divide the one from the other within the limits of the present notice. An attempt at such a definition as may convey what is required in an ordinary pony, so far as appearance is concerned, has therefore to be hazarded; and to meet this demand upon his powers of description, the writer may suggest that the best sort of pony for all-round purposes is a little animal as like a hunter as is possible, and as small as can be produced. This definition, at all events, possesses the merit of being subscribed to by more than one breeder of position, but it is not to be taken by readers as applying inflexibly to every little animal that professes to be a pony. It is more intended that it should be received as what the writer's idea is of the all-round animal, and not as a hard-and-fast description of what every pony should be like. The polo pony, for instance, should be simply a thoroughbred in miniature; whilst the weight carrier ought to be a dwarf Hackney, built on the most substantial lines, and boasting a supply of bone below the knee which would shock the susceptibilities of a hankerer after blood.

A pony, in brief, may with justice be described as a bantam of one or other of the larger breeds; as it is absolutely certain that any person possessed of leisure, patience and money, could, *in time*, effect a reduction in size, until pony dimensions were finally reached, of any variety of horse. In order to do so, many disappointments would have to be endured, and long years of anxious waiting for results would also have to be devoted to the task; but yet the end would surely come if the four great cardinal rules of pony breeding—viz., in-breeding, late foaling, short commons and an absence of pampering— were rigorously adhered to. It is not every one, however, who possesses the application or the means that will combine to enable him to apply himself with any prospect of success to such an undertaking, and therefore the value of a good pony is likely to be excessive for many a year to come, owing to the

infrequency with which real clinkers are met with, even by those who scour the country in search of a good-looking one that can go.

The production of a race of polo ponies, in particular, is one that appears likely to be opposed by many difficulties, the principal of which is the source from which the original stock would be derived. It is to the highest degree improbable, in fact, that the produce of thoroughbreds, who have for generations been kept for the greater part of their career in hot stables, would at first be robust enough in their constitutions to stand the effects of persistent in-breeding, and the hardships which alone can produce the article that is wanted. Yet, thanks to the truth of that admirable doctrine—the survival of the fittest—it is absolutely certain that time would effect that change in their systems which would enable them to found a strain of miniature thoroughbreds. At present there are plenty—although not enough—of clean bred ponies in the country, and it is to these that the founder of a polo stud would have to look for his foundation stock. It appears to be quite possible, moreover, that the introduction of a strong dash of Arab blood in the first instance would have the effect of strengthening the constitutions of the members of the stud; but once again, it may be repeated, in-breeding and short commons are absolutely essential to all prospects of ultimate success. If these methods are not drummed into the heads of would-be breeders, and if they fail to act rigorously upon these lines, it is almost certain that their time and money will alike be wasted. So far, therefore, it would appear that as regards the establishment of a breed of bantam thoroughbreds, the chief difficulty that persons who have set themselves the task will have to encounter will be the possible delicacy of their stock, for the lines and points of the clean bred article are so clearly established as to require that no experiments need be made to perpetuate the same.

It becomes a different, in fact an exactly opposite matter, however, when the foundation of a strain of blocky, weight-carrying ponies comes to be considered, as amongst these the constitutions would probably be vigorous enough, and it would be the type and tendency to run to size that would require the special energies of the breeder to be devoted to them. "Big little ones" can be produced with comparatively less trouble, though with no degree of certainty, from the crossing of a small animal with a big one; but this is a happy-go-lucky method of breeding, which only proves what everybody would expect—namely, that a small foal might be produced—and leads to no tangible results whatsoever. The youngster is, in fact, just as likely to come as big as its larger parent, but more probably will partake of the nature of both, and be a compromise between the pair, and therefore if by any principle of scientific breeding a strain could be produced which was in the habit of throwing cloddy, powerful young stock of about 14 hands in height, a certain fortune would be the reward of the happy founder thereof. Here, again, the advantages of persistent in-breeding would be fully recognised; but the first step on the path of progress must be the selection of the right sort of foundation stock. Fortunately, too, for those—if there are such in existence—who would be bold enough to attempt such an experiment, there is the Hackney at hand to commence upon, and this is the class of horse above all others which may be suggested for the purpose. The Hackney is, of course, a born mover, and most of them can walk, which in the case of an old man's pony is an essential accomplishment. There are big quarters, properly placed tails, flat bone and good feet, all ready to be utilised; and not the least advantage offered by the Hackney would be the hardiness of his constitution, which would certainly stand in-breeding and privation as well as could be expected or desired. Moreover, although the tendency of the age is to increase the size of the Hackney, there are plenty of, comparatively speaking, small animals at hand which, in spite of their breeding, are obtainable at fair prices. It also should be quite possible to obtain a few that were in-bred to commence with, and if so, the difficulties that presented themselves to a breeder would be smoothed away considerably at the commencement. At the same time there can be no disguising the fact that the establishment of a strain of thick-built ponies of a lower standard of height than that usually accepted for cobs, and which would breed fairly true, would be a very difficult task, as the earlier treatment of the foals would assuredly cause them to develop a tendency to lose substance, and if this were to be the case, the object of their breeder would be defeated.

Regarding what has been styled above the hunter type of pony, the magnitude of establishing a reliable strain would not be nearly so great, as there are in existence good-looking in-bred little horses which from time to time are to be picked up by those who keep a good look out for the same. It is therefore a source of surprise to many that in these days of agricultural depression, landowners and others have not devoted a portion of their energies to breeding ponies, utilising for the purpose some of the waste land that is on their hands. There is always a ready and fairly remunerative market for pony misfits, and as the cost of raising the animals is so very trifling, it is hard to see how the speculation could turn out otherwise than a very profitable one. Those embarking upon an enterprise of this description would, moreover, possess the additional satisfaction of knowing that they were doing good service to their country, as any movement that tended towards the suppression of the importation of the foreign element into British horse markets could not be regarded as being otherwise than patriotic. The country is now flooded by numbers of three-cornered looking ponies, with tails half way down between their croup and hocks, hailing from Russia and elsewhere, whose presence amongst us is only due to the apathy of English breeders, and does not therefore redound to the credit of the latter. The foreigners would never have invaded our shores had there been an adequate supply of home-bred ponies, and it certainly seems remarkable that, considering the national pride which most of us take in our horseflesh, the raising of ponies—the most profitable of all stock—should have been so completely discountenanced.

A short time ago the first volume of a Pony Stud Book was published, and before this volume has appeared there will probably be issued the initial volume of the Stud Book of the Polo Pony Stud Book Society. The future success of these registers will, of course, depend upon the methods in which they are conducted. A few breeders have for years past religiously preserved a record of their breeding operations, and were these pedigree books forthcoming, the basis of a Stud Book might be obtained. Under the happiest of circumstances, however, the formation of public records of the pedigrees of ponies, like the establishment of a reliable strain, must be regarded as a matter of time, patience and infinite research, for there can be no overlooking the fact that many of those who have bred the best of ponies have only succeeded by a fluke, and know next to nothing of the breeding of the animal which has made them famous. All this is far from being right, and it is therefore to be hoped that the future for British ponies—animals which can scarcely be put in wrong places if so be that their inches will not stand in the way—will be a brighter one than the past. At all events there is the satisfaction of knowing that a move is being made, and that the merits of the bantams have been clearly laid before the public. It will now be the fault of breeders—and certainly not of the ponies—if a perceptible improvement in the general quality of the latter is not evident before many years have passed over our heads. All that is necessary to ensure this is the devotion of patience and time to the improvement of a judiciously selected stud, by a man who has enough money to enable him to wait a few years. If these are forthcoming, and proper attention is devoted to the four cardinal rules of breeding alluded to above, there is little doubt as to the result.

There is a very considerable difficulty in existence now-a-days, in distinguishing one breed of pony from another, so far

Pony Stallion, Sir George 778.

at all events as the historical varieties are concerned; for so much foreign blood has been introduced into most of the most famous pony-breeding districts, that many of the original characteristics of each race have been either lost or improved out of recognition. Other instances exist where, owing to neglect, important features of a strain have been permitted to become absorbed by defects which, like ill weeds, will always grow apace. At the same time a few enthusiastic breeders—to their infinite credit be it said—have devoted their attention to the maintaining of a strain, and by their good offices many a pony lover is still enabled to indulge his tastes. Possibly, however, not one specimen in a hundred—assuming, of course, that it has not been raised by one of those owners who, having inherited a stud of ponies from his father, has permitted the stock to interbreed just as they pleased—can trace its pedigree back to the founders of the race, without encountering a bar sinister in the shape of a dash of foreign blood. In the case of ponies, as of many other kinds of stock, the periodical introduction of fresh blood is almost a necessary addition to the operations of a breeder, and certainly if the cross selected is the result of serious consideration, and its effects are closely watched—not for one generation, but for many—the experiment should be attended by success.

A possible increase of size may be apparent at first when a new strain is brought in, but the necessary, and in the writer's opinion, absolutely essential, amount of in-breeding being applied, this difficulty should be surmounted without much waste of time, or any great trouble on the part of the owner. In the case of well-established strains, such as that of Mr. C. W. Wilson, there should be no necessity for the introduction of any outside cross, as the shape and characteristics of the breed are perfect; but when it happens that a breeder has to commence the foundation of a strain with only crude materials to work upon, it is but natural that he should endeavour at once to gain possession of the required points and quality by the help of such animals as already possess the points he wants. It is useless to keep on in-breeding three-cornered looking wretches with each other. This can only have the effect of intensifying their faults, and making the eradication of the same more difficult when the time arrives for their effacement. In the case of good-looking little ones, the principle to act upon is entirely reversed, but, as has been observed in an earlier portion of this chapter, it should always be a case of sire upon daughter and son upon dam, and *not* brother upon sister.

Having thus expressed the difficulty that exists in discovering the identity of many ponies as they exist, the writer, with all diffidence, will attempt the task of presenting his readers with a brief outline of the chief points to be sought for in the most famous breeds. The selection of the order in which they shall be dealt with is in itself a delicate duty to have to perform, and, therefore, it may be best to deal with them alphabetically.

The Exmoor is unquestionably a very grand little animal for an ambitious breeder to commence working upon, always provided that he is fortunate enough to obtain possession of the right sort of materials. This variety is particularly hardy and sure-footed, averaging about twelve hands in height, and being usually of a good sound dark colour—principally bays, with mealy noses. Their heads are intelligent-looking and fine, with remarkably sharp ears, and their shoulders very good, which is a point that is very often deficient in a pony. Their backs are powerful, and legs short with good feet. The Exmoor is a most active, nippy pony, and is often a remarkable jumper for his inches, whilst his constitution is of the very best. That they afford a practical illustration of the possibility of in-breeding, combined with short commons, having the effect of dwarfing horse-flesh, is a fact that cannot be denied, and must be apparent to most persons who give their minds to a contemplation of the subject; but, unfortunately, there are very few of the old strain to be met with. About the time of the Battle of Waterloo, according to the assertions of

SHETLAND PONY, GOOD FRIDAY.
Winner of the Queen's Gold Medal, R.A.S.E., Windsor, 1889, &c.
The Property of Sir Walter Gilbey, Bart.

that not invariably veracious personage, "the oldest inhabitant," the breed was vastly improved by the fact that a foreign stallion, said to be the illustrious Katerfelto, was running on Exmoor; but since those far-off days, until recently, there has not been much new blood introduced into the district, and, unfortunately, but very few of the residents have attempted to keep up the breed in any definite shape.

Irish Ponies can scarcely be dealt with seriously as a distinct breed, for, although the writer would be the last to deny that plenty of good-looking and most serviceable little ones cross St. George's Channel every year, he is not aware of the existence of any particular characteristics of the Irish ponies, nor has he gathered any authentic information relating to the existence of a single particular strain amongst them.

New Forest Ponies have of late years received a great deal of attention from admirers of this class of animal, and a couple of years ago a society was formed for their special protection and improvement. There is very little doubt but that the original excellence of the race was largely due to the benefits received from the services of the Arab stallion who was introduced into the district over a hundred years ago, and who worked wonders when crossed with the original stock, which are stated to have flourished in the New Forest since the days of King Canute.

Shetland Ponies have become fashionable again, but the scarcity of good specimens and the extremely high prices which are set upon them, have placed these natty little horses beyond the reach of many who would like to own them. Possibly the Shetland is absolutely the purest-bred variety of horse in existence, that is to say, when he *is* pure bred, which is not invariably the case with animals passed off as such. His diminutive size, however, is so soon increased by the introduction of a cross that the detection of foreign blood is not often a matter of much difficulty, and certainly the majority that can be picked up in the island are uncontaminated by such a taint. The Marquis of Londonderry's strain is admittedly the best in existence, but even his experienced manager finds it tax his energies to the utmost to keep the size down, as Master Shetland is disposed to grow too big when indulged in respect of good feeding and housing. The head of a Shetland is very delicate and fine, even for an animal of his inches; his short neck is small at the setting on of the head, but thickens considerably at the shoulders, which are usually short and straight. His back is short, ribs well sprung, and quarters very big compared to his size, whilst his legs are flat and feet rather round. Ten hands or a little under is the average height, but smaller specimens are occasionally met with, and when they are they realise a good deal of money if well made. The prevailing colours are bay, brown and dun, but occasionally a black or skewbald are met with, and very rarely indeed a white.

Welsh Ponies have earned a most enviable notoriety for all-round excellence, but this is scarcely to be wondered at when it is remembered that their height varies so considerably that a limit of nearly three hands is permissible among them. Doubtless a good deal of new blood has been introduced to these denizens of the Principality, but the value of the old strain is clearly proved by the fact that, cross him as you may, the Welshman will always stamp a heap of the pony about his descendants. Varying as they do so much in height, it is a very difficult matter with most men to be asked to pick out a Welsh pony from a drove of which he knows nothing, but a prominent eye is a peculiar characteristic of the race, and this point is one that is regarded by many as an index to the breeding of its possessor. They are usually very good in legs and feet, but this is scarcely a remarkable feature in a breed which has the advantage of being for the most part raised in a mountainous country, where surefootedness is absolutely indispensable to all animals inhabiting the same.

Welsh Pony, Tommy.
The Property of the Irish Congested Districts Board.

CHAPTER IX.

ASSES AND MULES.

ALTHOUGH in this country asses and mules do not occupy so important a place as they do elsewhere, their numbers are not inconsiderable, and as a good many mules are used for light draught, we may give a few notes here in reference to both varieties. The services of the ass are with us chiefly utilised by costermongers, gipsies, small tradesmen and hawkers; they are also used for carrying young children, and for seashore riding; but the animals employed for these purposes are, comparatively, of a small type. Very different are the asses of France, Tuscany, Spain, Persia, Asia Minor, Arabia and Egypt, which are of large size, and a big variety is also maintained in the United States of America, and in South America, for mule-breeding, which in various parts of the world is an important industry.

One of the best breeds of asses is to be found in the Poitou district of France, and of these an excellent description was given by Mr. Charles L. Sutherland, in his report to the Richmond Commission on Agriculture. From this we quote the following:—" Poitou is the French breeding ground and nursery of the heavy draught mule. The Poitevin mule is the best mule for farm work, and a good specimen is very nearly as big and as heavy as an ordinary cart-horse. His peculiarities are that he is short-legged, short-jointed, and big barrelled, with great knees and hocks, and plenty of bone below the knee, while his feet are comparatively large; and less contracted than those of other mules. These peculiarities are derived from the Poitevin jackass, a variety as curious and perhaps as ugly as he is massive, short-legged and valuable, and one in which the Darwinian theory of selection has been worked out in its entirety. The mules are worked on the farms from the time they are eighteen months old, till they reach three or four years, when they are sold to dealers for the 'Midi,' Spain, Italy, &c. They vary in height, from 15 hands to 16.2, on short legs, and a good 16 hands four-year-old mule is worth from £60 to £80. Inferior animals may be purchased at from £30 to £40 each. The chief fairs are held in the months of January and February, but most of the good animals will have been previously bought privately at the farms, such is the demand that exists for them. A man may buy fifty mules in St. Louis, in the United States, in the same time that it will take him to buy a single mule on a Poitevin farm where bargaining is carried on *ad nauseam*. The jackass, or *baudet*, is the most important of all quadrupeds in Poitou. He is the sire of the mules, and as such is the direct means of putting large sums of money into the pockets of the farmers. The price of a young improved animal of two years varies from £80 to £120; a good proved mule getter, four years old, from 14 to 15 hands high, is worth from £200 to £320, and one was sold in the Vendée, just before the Franco-Prussian war, for £400. These valuable animals are kept in a filthy state, are never groomed, and never taken out of the building in which they are kept, except perhaps to be shown to a visitor or possible purchaser. The fee for the service of each mare is from 16s. to 20s. The female asses are rarely parted with, except for some defect. Their value may be set down at between £24 and £40. The *Conseil Général* of the Deux Sèvres votes annually the sum of £200 for prizes for mules and asses at the local shows. These establishments are technically called ' *ateliers*,' and the fact of owning such an establishment entitles the proprietor to the right to call himself ' *Maitre*,' and gives him a position in the country. Each stud farm consists of from four to seven stallion asses, a stallion horse, a 'teaser,' and one or more she-asses. The mares are always brought to the stud farms, of which there are 160 in Poitou, the Deux Sèvres alone claiming 94, with 465 jackasses." The dimensions of a good fair specimen of the Poitou ass, suitable for breeding heavy draught mules from cart mares, are as follows:—Height 14.1, forearm 19½ inches, knee 15 inches, below knee 8½ inches, hock 17½ inches, below hock 12 inches, greatest girth 77 inches, girth behind shoulder 66 inches, length of head 25 inches, length of ear 15 inches, ears, tip to tip across, 32 inches. The Poitou breeders always select black or brown bay donkey sires with white bellies, and will not have greys. The tip of the nose must be of a greyish white, and covered with a slight down.

Maltese asses have also been for many years highly prized for breeding mules from blood mares both in the East and West Indies, and they have consequently become very scarce in Malta and Gozo. Sir Robert Biddulph, when Governor of Cyprus, reported so favourably of the Cyprus donkey as possessing every characteristic of the Maltese, that he was directed to supply a certain number to the Government of Bombay for mule breeding. The exportation proved perfectly successful, and, on quitting his governorship, Sir Robert made over to his cousin, Mr. Ralph Palmer, at Nazeing, Essex, a jack and two jennies which he had kept in Cyprus for his own use. Captain Fawkes describes the Maltese ass as follows:— "A pure-bred Maltese ass has a perfectly black body, neck and legs, with white or light grey under the belly and inside of legs. His head should be light, with active ears, and eyes fringed with tan and white; as also his muzzle, which gives a nice expression to the animal's countenance. The height is about 13.2, and the girth 5 ft. 3 in." Of the Cyprus ass, General Sir R. Biddulph says:—"As regards pedigree, the Cyprus asses are said to be of Syrian blood; the grey to be a cross between the white ass of Damascus and the black ass of Syria. The Cyprus asses are excellent animals to carry heavy loads and are very good-tempered."

Speaking generally of the ass, it may be said that he possesses unusual hardiness of constitution, and is capable of enduring great fatigue, while he is sure-footed and little liable to disease. Dr. Fleming, C.B., in the "Practical Horsekeeper" (Upcott Gill), says, "In no other animal, perhaps, is good feeding, kindness, grooming and housing more amply compensated for by increased service and willing performance than with the ass. His appetite is not large, and he is much less fastidious about the quality of his food than the horse. A few pounds of hay and oats in the course of the day and night will maintain him in excellent condition, and even on hay or grass alone he will perform a fair amount of work, but if the toil is exacting, the food should be in proportion. A large-sized ass will get through a wonderful amount of work on half-a-dozen pounds of oats and eight or ten pounds of hay. The limbs of the ass should be strong; the knees and hocks large and free from blemishes; the feet not too small, and the hoofs sound; the chest wide; the back unscarred; the body rather long, but compact; the hind quarters and croup round and wide." The period of gestation with the she ass is twelve months. The foal is weaned at nine months. The ass should not be put to hard work under four years of age.

The mule is a hybrid, the resulting produce of a cross between the male ass and the mare horse; the result of the opposite cross, between the horse-stallion and the mare ass, being the hinny. Although, as we have said, there is little mule-breeding carried out in this country, the number of mules in the United States is 2,314,000, while in France there are 227,000 mules and 361,000 asses. Mules are invaluable as beasts of draught and beasts of burden in the South of Europe and in certain parts of Northern and Southern America. Dr. Fleming says the mule is one of the very best beasts of burden man possesses, and for this purpose he is employed chiefly in mountainous countries and those in which wheel-carriage cannot be resorted to. He is also greatly in request for transport purposes during war, his

patience, robustness, and endurance of hardship and fatigue rendering him particularly well-adapted for the exigencies of field service. It is claimed that he is much stronger than the ass, more capable of bearing fatigue than the horse, less restive under the pressure of heavy weights on his back, and his skin being harder and less sensitive, renders him capable of resisting better the sun and rain. He lives as long as the horse, costs less, is more suitable as a beast of burden, and far superior in surefootedness. Mares, 14 to $14\frac{1}{2}$ hands high, put to the largest donkeys, produce good mules for draught or saddle; for pack, the best size for mares is between 13 and 14 hands. The most convenient height for mules is from 13 to $15\frac{1}{2}$ hands, the average being 14 to 15 hands. A mule is scarcely full grown at five years old, and is fit for full work at six to seven. Mr. John Thompson, agent to the Duke of Beaufort, at Badminton, wrote as follows to the author of the " Book of the Horse ":—
" Mules were first introduced to Badminton about seventy years since. The first Spanish jack was imported during the Peninsular War, and the first mules by him were out of a large active cart mare. Three or four which she bred were upwards of $17\frac{1}{2}$ hands high. Mule teams have been kept up ever since, chiefly home bred; and, in consequence of the difficulty in procuring first-class jacks, imported animals have latterly been introduced. We have bred them from both cart and half-bred mares, and find that the stock from these are more powerful than the imported animals, being larger in the bone and of greater substance. We have had jacks from Malta and Spain, but those from the latter country are generally superior. The mule foals are very hardy, there being no difficulty in rearing them, and, when grown up, they are less expensive to keep than horses. Ordinary carters drive the teams, which are composed of four mules each driven double, and they will each with ease draw a load of 50 cwt., in addition to the waggon, at the rate of four miles an hour on a good road. They are especially useful in carrying hay or corn

during harvest, being much quicker than horses with light loads. They last longer than horses, a mule at thirty years being about equal to a horse at twenty." Mr. C. L. Sutherland, Down Hall, Farnborough, Kent, worked a farm of 300 acres (90 being arable), at Coombe, Croydon, entirely with mules, which consumed a bushel of oats each per week, with green clover in summer, and one and a-half bushels of oats each per week, with hay, during winter. It is very important that mules be driven by those who understand their peculiarities and who can properly manage them. A mule measuring $14\frac{1}{2}$ hands high, should weigh not more than 1,000 lbs. In the United States, mules in towns generally receive a mixture of maize and oats in the proportion of 1 to $1\frac{1}{2}$, the quantity of the mixture allowed per diem varying according to the size of the animals from 4 to 10 lbs. Together with this grain ration, from 6 to 12 lbs. of hay is given.

CHAPTER X.

MANAGEMENT OF LIGHT HORSES.

Stables.

THERE is a right and a wrong way of building both a stable and a house; but the ideal in each case has often to give way to the necessities of position and the shape of the ground available for building purposes. In the country where land is comparatively cheap, and space not of much account, very many perfect ranges of stabling are to be found; but in London and other large towns the most valuable horses are frequently housed in most unhealthy habitations; and the wonder is that the veterinary surgeon is not in even greater request than he is. Some of the cab stables are as bad as they can be; but the salvation of the cab horse is that he spends so much of his time in the open air, thereby counteracting the pernicious effects of his close dwelling; where, however, improvement is not possible, suggestions are of no use. It is often necessary to pack away the greatest number of horses in the least possible space; and when this takes the form of a long stable with a door—often the only means of ventilation—at one end, it will be readily understood that the unlucky steeds at the far end get no air to speak of—certainly no fresh air.

So far as is possible, stables should be both light and airy. Sunlight is beneficial to both men and animals; and a horse brought out of a dark dungeon sort of place is very apt to shy and become frightened. Moreover, where there is darkness there is generally dirt, for a stableman cannot see to thoroughly cleanse the place, and unless a stable be scrupulously clean it cannot be healthy, while the slightest particle of stale food left in the manger will often cause a horse to refuse his corn. Nevertheless, we know that at any rate in towns, it is impossible to have stables as they should be, and in all probability both cab horses and expensive carriage horses will continue to be housed in places which are admittedly unfit for equine habitation. When about to rent stabling, however, the horse-owner would do well to decline at any price stables which are dark, or which have no other means of ventilation than a door or window at one end, if, that is to say, the stable contains more than about two stalls. We may, however, remark in passing that by light stables we do not mean glaring ones, and no horse should be housed in a light stable, the walls of which are whitewashed all round. In stalls the wall above the manger to rather more than the height of a horse's head, should be of a cool, neutral tint colour; and in loose boxes the same arrangement may prevail all round.

It is scarcely within the scope of this work to treat of the necessity for adequate ventilation from the scientific point of view; but the reader may be reminded that without a sufficient supply of fresh air the blood cannot be in a proper state; and when the horse breathes, whatever there is impure in the blood is given off into the air, so that if there be no adequate ventilation the horse breathes again the impure air, consequently the purity of the blood and the general health of the horse depend greatly upon the quality of the air inhaled into the lungs.

In that most excellent book which should be in the library of every horse-owner, " Horses and Stables," by Sir F. Fitzwygram, the author says: "Fortunately the peculiar properties, or rather the state of the gases which respectively constitute foul and pure air, afford great facilities for ventilation. Heat causes all matters to expand, some more and some less; but gases under the influence of heat expand very rapidly, and to a very great degree, and as they expand they of course become lighter."

As a general rule, foul air in a stable is also heated air. It is only necessary to breathe on the hand to feel that our breath is warmer than the air. The foul air being lighter than the pure air, the former rises, and this of course suggests that the most efficient method of getting rid of it is to provide an outlet at the top of the stable where the foul air will collect. In principle, therefore, it is easy enough to ventilate a stable, but in practice difficulties often present themselves. Some stables are low and have lofts over them, and when this is the case the employment of louvre boards—the best means of ventilation—is impossible. Amateur ventilation often takes the form of knocking holes in the walls here and there, with the result that the horse stands in a perpetual draught. The owner knows that ingress must be provided for the fresh air, and an exit for the foul air; but in making one and the other a draught is created, and this will also be the case when there is also ventilation from above, for as soon as the foul air has escaped the pure air will come in at the openings and draw down on the unlucky horses below.

In order, therefore, to get a properly ventilated stable it is necessary to have apertures at the roof to allow the foul air to escape, and openings lower down to permit of the ingress of fresh air, and how to do this without at the same time creating a draught is one of the difficulties against which he who constructs a stable has to contend; it is far easier to provide for the escape of the foul air than for the ingress of the fresh supply. With the aid of professional advice it is a comparatively simple matter to have a properly ventilated stable where space allows of the length, breadth, and height being just what they should be; but, when buildings of all sorts of shapes and plans have to be occupied, the difficulties are great enough to baffle even the most learned in sanitary engineering. At any rate anything is better than a stuffy stable, and if the worst comes to the worst it is better that doors and windows should be left open, even if a canvas screen has to be used to protect the horses nearest to the opening, or if they have to be additionally clothed, than that they should constantly have to breathe foul air. When the horses are at work common-sense suggests that doors should be left open, and especially when the stabling consists of single loose boxes which do not communicate with each other. These are frequently defective in ventilation, and windows and doors are the only means whereby places can be kept sweet. Stables built up against a wall are the most difficult to ventilate properly, because no ingress of fresh air can be provided through the external wall; otherwise, a simple plan is to have a perforated brick inserted below the manger. This will supply each horse with a fair quantity of fresh air, and if the architectural peculiarities of the building allow of an escape for foul air at the top, the main requirements of ventilation will have been complied with.

Paving.

A good system of paving and draining is indispensable, and if either or both are defective a sweet stable must not be expected.

As a material for a floor we want something that will not absorb the urine, that will wear a reasonably long time, and that will afford a good foothold to horses. Each of these requisites is important, but in the opinion of the writer too much is often sacrificed to foothold. In order to gain this, channels are cut this way and that, the edges of bricks are bevelled away, so that when two are laid alongside of each other a channel is formed. So long as they run longitudinally there may not be much harm in these grooves, but it must be remembered that every inequality in the surface holds urine and particles of litter, and these tend to make a stable smell. On the whole the writer is inclined to advocate a paving of the very best yellow clinker bricks, and with no more slope either to the centre or fore and aft than is absolutely necessary to cause the water to run off. The bricks should, of course, be laid in cement about four or five inches deep. The channels are best left alone as the litter will give the horse a foothold.

Draining.

We learn from the Badminton Library that the Duke of Beaufort solves the difficult question of how to drain stables by having no drains at all. Each loose box is paved with stone slabs, and there is no drain whatever, the moisture being absorbed by the straw. There is no doubt a good deal to be said in favour of this plan, inasmuch as the smell of a stable arises not so much from what has been freshly dropped as from the remains of previous droppings, and the odour caused by faulty drains, those which permit of the smell to travel from the sort of cistern sometimes used. The writer speaks feelingly on this subject, as a few years ago he took a house to which was attached some apparently excellent stabling. On opening the doors of the loose boxes the smell was, to use a common phase, "enough to knock you down." A workman was sent for, and it was then discovered that the urine from each of the boxes drained into a tank which was connected with the drain pipe. In this tank was found the deposit of years, and the only wonder was how the horses of the previous tenant had managed to exist at all.

The malodorous state of the above mentioned stables evidently arose from a desire to keep the urine for manuring purposes—an object laudable in itself, yet as a rule wholly incompatible with having stables in a sweet and wholesome condition, unless the connection between the tank and the pipe is effectually cut off. In draining, as in ventilating, position may enter largely into the question; and a plan which may be quite feasible in one situation may be impracticable in another. Generally speaking, however, surface drains, as they are called, are the best, because they are more easily kept clean; if drains run underground anywhere near a stable there is always a very great chance of foulness. In consequence of the high price of straw, peat moss litter is a good deal used both in private and trade stables; but whatever may be its merits in other ways—and it has several—it has a tendency to choke the drains. This is a somewhat serious matter when stables are drained in the ordinary way, with a drain in the middle of the box or stall, and where peat moss is used it will, in the long run, be found better to cover the drain with some contrivance which will allow of its removal when required, so that the drains may be flushed.

At the same time it is only a small number of horse owners who are in a position to have their stables exactly as they wish. Consequently, in spite of all theory on the subject, the horse owner will probably find himself in possession of a stable with underground drains. When this is the case it is absolutely necessary that the drains should be properly trapped; the gratings in the centre of the stall or loose box must be lifted *every* day, all scraps of litter must be taken out, and the drains flushed regularly. This indispensable proceeding is of a somewhat unsavoury nature, and as it gives no visible results is often shirked. Fortunately the nose is a very good guide on entering a stable from the fresh air, and should a pungent smell of ammonia greet the visitor he would do well to question his groom about the drains. A good deal more might be written on the subject of drainage, but the object of these pages is merely to put the horse-owner on his guard. No directions would suffice to enable any one previously unacquainted with the details of draining to devise a system for his own establishment; the most effectual plan, therefore, is to call in the assistance of an expert, just as one would in connection with the drains of a house.

There are in existence two bodies professing themselves ready to give advice and aid in the matter of drains. One is the "Sanitary Security Association" (1, Mitre Court, Fleet Street, London, E.C.), the other the "North Eastern Sanitary Inspection Association," which has Sir Matthew White Ridley for its President. The addresses of the latter are 4, Chapel Walk, Cross Street, Manchester, and (Head Office), Neville Street, Newcastle-on-Tyne. Possibly one or both of these Societies may keep an expert in stable drains, but the reader may be strongly advised, when seeking professional advice, to have recourse to some one who has made a special study of stables. Meantime he may read with much profit to himself the sections dealing with stables in that excellent book "Horses and Stables," by Sir F. Fitzwygram (Routledge), and a book on stables from the pen of Mr. John Birch, the architect (Blackwood). As already mentioned, there may be several reasons why it is impossible to render stables perfect; but, if proper ventilation and drainage can be secured at an outlay within the means of the owner or tenant, it will be money well expended.

The number of cubic feet of air for each horse is a matter of some importance; but Sir F. Fitzwygram tells us that it is not known for certain what is the minimum quantity of air required for each horse, but possibly about 1,200 cubic feet would be sufficient.

Stable Fittings.

Loose boxes are always to be preferred to stalls for all horses, but it is not always that they can be provided. Stalls should have a minimum width of 6 feet, and should be $10\frac{1}{2}$ feet deep. Horses like company, so the partitions should not be high enough to prevent them seeing one another, though on the top of the wooden partitions there may be some open iron work. This arrangement should be adopted in all ranges of stabling; and in the writer's opinion the very worst form of stabling is where each loose box is unconnected with any other, and is simply a cell for the solitary confinement of each horse.

Doors sometimes work on runners so as to slide—an arrangement that is sometimes necessary owing to circumscribed space. There is, however, no objection to a common hinged door, which of course should open outwards. Kay's locks will be found convenient for stables, as they cannot be opened by the horse, and they require no slamming. Stable doors must be both wide and high—four feet wide at least, and eight feet high. These measurements will guard against the harness or hip bone from being caught by the door posts, provided, of course, that the horse be led in and out carefully, while if he throw up his head he will not strike it against the top—a mishap that sometimes makes it difficult to get a horse in and out of his stable.

The ordinary hemp halter should never be seen in a private stable. All the horses should wear leather head collars, and the rope—chains are noisy and leather straps get gnawed—should not be longer than is needed to allow of the horse lying down in comfort. The log should be sufficiently heavy to keep the rope taut. There are one or two contrivances in which a spring is used instead of a log; but they sometimes lead to the use of a leather strap for the purpose of easier winding.

Mangers, &c., can be had at prices to suit all classes of buyers; but the objection to wooden mangers is that particles of food can lodge in the angles at the ends and sides, whereas with those which are round it is impossible. On the inside of the manger there should be a slight rim to diminish the chance of the corn being thrown out. There are several firms which make iron stable fittings; and from the show rooms the horse-owner may select everything he wants. He will find that the hay rack is, in these iron fittings, on a level with the manger—the proper place for it; the old plan of placing it high up is most absurd, for not only is the unlucky horse compelled to assume an uncomfortable attitude when eating his hay, but the seeds and dust necessarily come into his eyes.

Stable Management.

Having made the stable as complete as possible, the next thing is to adopt a judicious course of stable management, in which of course the groom will play a very prominent part. Some persons are fond of saying that they are their own stud grooms, but practically the groom does almost what he likes. As he has to deal with valuable property it is advisable to engage as good a man as can be procured; and if one can be thoroughly recommended it is worth while to give an extra shilling or two in wages. A man who has been trained under a good stud groom is always to be preferred to one who has picked up his knowledge anyhow; but stable servants of all kinds are desperately conservative in their notions, and can hardly be prevailed upon to make any change from what they have been taught.

Feeding.—Every horse must be fed with reference to his size, his natural appetite, and the work he has to do. The 16 stone hunter needs a more liberal dietary than the 15 hand hack.

There are no horses over 15 hands which should have less than three-quarters of corn per day, and this will be about the right quantity for horses that do easy work in the park, and it may suffice to feed them three times a day; but the writer has a preference for feeding four times a day. Hunters will always require at least four quarters of corn a day, while weight-carriers and those which hunt twice a week will need five or six quarterns. A quartern is the fourth part of a peck; and, when oats weigh 40 lbs to the bushel, weighs $2\frac{1}{2}$ lbs. Beans should be given to hard working horses alone, and to no horses until they have turned five years old. Beans should be old, split, and a double handful twice a day will be plenty.

Opinions vary as to the best method of giving hay—a very

necessary article of stable diet, inasmuch as a horse's stomach needs mechanical distension. Some people give no chaff at all with the oats, others give no more than a handful or two, while others, again, give almost all the hay in the form of chaff. When long hay is given there is nearly always a great deal of waste, as horses pull it out of the rack, let it fall on the floor, and trample it under foot; in my opinion, therefore, it is the better plan to give nearly all the hay in the form of chaff, giving no more than a little long hay the first thing in the morning and the last thing at night.

The quantity of hay to be given will vary from one to two trusses, that is to say, from 56 to 112 lbs. One truss, which gives a daily allowance of 8 lbs., is the very least that can be given to horses of 15 hands and under; for an ordinary park hack about 10 lbs. may be set down as the day's portion. For weight-carrying hunters and full-sized harness horses two trusses a week will scarcely be too much; though in some hunting stables it is the custom never to give more than a truss and a-half a week, but to give a few extra oats; but hunters able to carry very heavy men will want not less than two trusses of hay per week, no matter how many oats they have.

Some people give a little bran with the food twice a day or oftener, but the utility of the practice may be doubted; bran in a dry state is an astringent, and therefore not required with the majority of horses. In the form of a mash it is a laxative, and a bran mash should be given at least once a week to counteract the feverish symptoms which are supposed to result from high feeding, but which, it is submitted, need not follow if a proper system of feeding be adopted. To the dietary of hay, oats, beans and bran mash, some carrots should be added in season, and a little green meat in the summer; but very much depends upon watering, and in this connection grooms have much to learn.

It is commonly said that unless a horse be in a state bordering on high fever he cannot be in high condition, and in pursuance of this theory, horses are stuffed with stimulating food, and then physic is administered to cool them down. It is submitted that any system of feeding and stable management which engenders this fever, and demands physic to check it, is wrong, at any rate in the case of horses which are wanted, not on a particular occasion like a race horse, but for two days a week or three days a fortnight, like a hunter.

It may be admitted that there cannot be much variation in the matter of food. Hay, oats and beans must form the stable diet of saddle and harness horses, and bran mashes, carrots, and green food can only come in as alteratives. It is true that wheat, barley and some kinds of prepared food are sometimes recommended; but, on the whole, it will be found expedient to confine the articles of food to those which have stood the test of time, especially when the horses are hunters, or harness horses whose work is hard and fast. It is, of course, well known that the London General Omnibus Company feed largely on maize, and for horses doing slow work it answers fairly well when mixed with oats, but it is too heating and fattening for horses doing fast or very easy work. New oats should not be given to horses. They often cause horses to scour, and to sweat with the least exertion. Why this effect should be produced by new oats is not, so far as the writer is aware, known; but the fact is plain enough. By Christmas, however, a change comes over the oats which have been cut during a preceding harvest, and after Christmas those which have been "well got" are permitted in many hunting stables, though it is better if possible not to use oats till they are one year old; that is to say, oats harvested in 1893 should not be used till the autumn of 1894. Some people prefer white oats, others pin their faith on black ones; but so long as they are thin-skinned, full of husk, and weigh at least thirty-eight pounds to the fair bushel, it does not matter whether they be black or white; but palpably yellow oats should not be chosen when the others are available.

Almost of more importance than food to the hard-working horse is a proper system of watering; and the writer ventures with some confidence to express his conviction that if grooms paid more attention to this important detail of stable management, feverish symptoms would not be nearly so common as they are, nor would the dose of physic be in such constant request. Save on one or two occasions to be mentioned in due course, horses should be allowed to drink as much as they please, and this, if they are not kept without water too long, will never be of any great amount, for horses drink no more than nature requires. But, as a horse has no means of knowing what work is to be required of him, the groom's common-sense must step in if the horse be wanted for fast work soon after stable hours, and the usual quantity must then be diminished. On hunting mornings, especially if the horse have some distance to travel to the covert side, there is not the least reason to stint the animal to any great extent, or, as is sometimes done, to deprive him of water altogether. This is a most cruel practice, and is based merely upon the prejudice of the groom. If experiments were tried it would be found that no hunter would be one whit the worse for a reasonable amount of water on hunting mornings. During the hunting season hounds never meet before half-past ten, so something like three hours would elapse between the consumption of the water and the commencement of the work, and this is surely long enough to enable him to get rid of a moderate amount.

After hunting it rests with the master himself to take the first step towards comforting his horse; and no time should be lost in giving him half a bucket of gruel, or failing that, of chilled water. A cake of Rumney's food carried in the waistcoat pocket renders the rider independent of the meal or flour of the inn, and enables him to give his horse a bucket of nourishing gruel whenever a little hot water is obtainable. On arriving home he will have more gruel, and before he is done up for the night the hunter should have thoroughly quenched his thirst. Horses should always be watered *before* they are fed, as owing to their internal construction there is some danger of working some of the oats into the gut if the process be reversed.

The best possible arrangement is to let horses have water always before them. The receptacle containing it can easily be fitted with a cover; and it will be found that horses so supplied rarely take more than a sip or two at a time, and so are never unfit for moderate work at a moment's notice. The cover can be put on the trough before a horse goes out hunting, and when he returns hot and tired; but at other times the horse can be left to his own devices, and he will take no harm.

The idea that a horse must subsist during the hunting season on dry stimulating food with the minimum of water has no consideration, scientific or otherwise, to support it, and a hunter can no more be at his best if stinted of water than he can on a short allowance of oats. Unless solid and liquid are consumed in due proportion mischief is sure to result, and when too little water is given feverish symptoms are at once engendered, and doses of physic are required. Any groom can, if he choose, test for himself whether it is advisable to start a horse for a long and heavy day's work either without any water at all or with an insufficient allowance. Foot beagles are now to be found in nearly every country, and let a groom ask for a day off, and have a day's running with them, abjuring all liquid for his breakfast. If he will also carry out another fad, viz., not to allow hunters to take anything out of their own stables, he will be in a position to form some opinion as to the merits of the system. Seldom is the groom found to refuse a proffered glass of beer, be the occasion what it may; yet many are found to subject their horses to a thirst which they themselves would not endure for five minutes.

Grooming is a very important part of stable management, since it is to the skin of the horse what washing is to the human skin. It is a mistake to wash either the body or legs of a horse; it does no good, and takes up a great deal of time, as after washing the horse must be rubbed till the coat is perfectly dry. Good strapping, however, cleanses the skin by clearing off what has exuded from the pores, and conduces to a good coat, and it may here be remarked that no groom should in any circumstances be permitted to give any drug with the idea of producing a glossy coat. The curry-comb, it need hardly be said, should never be used to a hunter or harness horse; its use now is to cleanse the brush from the scurf coming from the body. In order that time may not be wasted, and that the horse may not suffer by make-shifts, there should be as many sets of "tools" in use as there are helpers, so that each man may have a set to himself.

Clothing is a matter of individual taste. The best kersey with the owner's initials on it looks very nice and smart; but for all practical purposes the fawn rug is just as good; while for night wear, the outside rug may be of jute lined; these rugs are very cheap and useful. When a horse is at work the opportunity should be taken of exposing his rugs to the air, instead of leaving them huddled up in the manger as careless servants will sometimes do, and they should be periodically beaten and brushed. In short, extreme cleanliness is a *sine quâ non* in all departments of the stable.

In many establishments there is a perpetual battle going on between hot stables and light clothing, and cool stables and heavy clothing. Both arrangements have their advocates just as both have their weak points. In this matter, however, as in others, extremes should be avoided. It is useless to say that a stable should be kept up to a certain temperature, because during the continuance of a hard frost the stable may be many degrees below the ideal temperature; and when that is the case the owner need not fear to put on loose bandages, three rugs and even the hood; but speaking generally, a couple of rugs and a moderately warm stable will best preserve the horse's health.

Exercise is the last, though by no means the least of the things necessary to keep a horse in health. The amount of exercise will, of course, depend upon the amount of work the horse is required to perform. The doctor's horse which daily goes his rounds will need none at all; whereas, the hack used on fine days for an hour and a-half or in the park requires a good deal, say two hours every morning, and ladies' horses should always have their backs kept down by plenty of work. Some grooms always walk their horses at exercise; but this is a mistake, as two hours at one pace is apt to produce a certain amount of leg weariness, so spells of slow trotting should be introduced. Nor is it advisable always to exercise horses on the flat. A moderately hilly route should be selected if possible. In the case of small studs cantering or galloping at exercise will seldom or never be necessary. With an overgrown stud, on the other hand, the members of which come out but comparatively seldom, a sharp canter or two as the day for the horse's "turn" draws near, may be expedient; but as a well paid and presumably experienced stud groom will have charge of these large studs, nothing need be said here as to their management. No horse, however, whatever his value, should be exercised without kneecaps.

Feeding and Watering.

The following more detailed notes on feeding and watering horses are by Dr. George Fleming, C.B., F.R.C.V.S. They appeared originally in the *Live Stock Journal Almanac*:—

"The subject of horse-feeding is one which should interest every horse-owner and horse attendant, as upon the manner in which horses are fed will greatly depend their health and usefulness. The judicious feeding of all animals domesticated by man assumes considerable importance from an economical point of view; but with the horse it is a matter of the greatest consideration, and for more reasons than apply to the other creatures which minister to our comfort, convenience, and pleasure.

"In the first place, it is his strength and speed which render him so valuable to man, and to ensure his developing these qualities to the greatest advantage the food must be different in quality to that given to ruminants; while the mode of feeding should also be adjusted to his anatomical and physiological peculiarities. His stomach is small and only capable of containing about three or four gallons of fluid, whereas that of the ox may hold from twenty to twenty-two gallons.

"The horse must, therefore, receive food more frequently, and, consequently, in smaller quantities at a time, than ruminants, in order that he may digest properly, and without injury to the stomach and other organs.

"The chief or typical food of the horse, in this country at any rate, is hay and oats, and of the two, perhaps, the hay is more essential to maintain life and health under ordinary circumstances, when unusual exertion is not demanded. Horses can live altogether on hay provided they are not called upon to do any, or at least much, labour, and it is therefore designated the material for supplying the internal or vital work of the body; but if the muscular system is called upon for unwonted exertion, then more nutritive food, in smaller bulk, must be given to enable the body to perform what may be called external work. Horses can therefore live and thrive upon hay or grass alone, and even do a certain amount of slow work—but then, a large quantity is needed. For instance, for a moderate-sized horse from eighteen to twenty pounds of hay, or even more, are needed as essential diet—that is, to perform the internal work of the body—for twenty-four hours.

"So long as a horse has plenty of time to eat it, a hay diet causes no inconvenience because of its bulk, but when leisure is not allowed, or when he has to exert himself after a sufficient meal of hay, then injury is likely to be done.

It has been remarked that, in the process of mastication, dry hay becomes mixed with four times its weight of saliva, while oats only require an amount of saliva equivalent to their own weight. It is said that a horse, in eating ten pounds of hay, loads his stomach with forty pounds of saliva in addition, or fifty pounds in all; but in consuming an equivalent amount of oats, say five pounds, he needs but five pounds of saliva, or ten pounds altogether. Therefore, in introducing into the system a given amount of flesh-forming aliment in the form of oats, the stomach is filled to only one-fifth the extent that would be necessary if the same quantity of nutrient material were given in the form of hay. But the ten pounds of hay, with its saliva, could not all be accommodated in the stomach at once, but only at three times, unless the organ is to be distended to more than the normal plenitude of two-thirds of its full capacity. The five pounds of oats, on the other hand, with its five pounds of saliva, will not fill the stomach to one-third of its capacity, but leaves the amplest opportunity for freedom of movement and the secretion of the gastric juices. It might be added that the ready digestibility and assimilation of the food is a very important matter, and especially for hard-working horses, with which time is all-important. If the food is difficult of mastication, and requires a lengthy period to reduce it to the necessary condition of crushing and insalivation before being swallowed, then the animal gets less rest, and so much power is wasted by the muscular movement of the jaws; while, if it is indigestible, it takes a longer time to reach the blood, and fatigues the stomach before it is in a fit state to enter that vital fluid, besides loading the bowels with matters which are often worse than useless.

"If we attempt to feed a working horse upon hay alone, we must give him much more than if he were not doing any labour; if, for instance, he is allowed twenty pounds per day, as has been observed by the American *Live Stock Journal*, he will require four or five hours in order to masticate it properly; and if this quantity of hay must be saturated with four times its weight of saliva, so making a total of one hundred pounds, the stomach will be filled to five or six times of the physiological condition of two-thirds of its capacity. For really hard work the ration would have to be doubled; but if it be increased to thirty pounds only, the period of mastication will be prolonged to at least six hours daily, and the mass of material swallowed will amount to one hundred and fifty pounds, which will fill the stomach eight times in succession. This state of things is manifestly incompatible with hard or prolonged work, to say nothing of the over-distended belly, the impaired wind, and the soft flabby muscles, which unfit the horse for anything faster than a walking pace, or severe exertion even at that. One of the chief causes of one form of broken wind, which consists of rupture of the air cells of the lungs, is working horses severely while their stomach and bowels are distended with bulky food.

"For these reasons, then, horses which are required to work for long periods, or to get through a large amount of exertion in a comparatively short time, should have their food presented to them in the best form possible with regard to mastication, digestion, and assimilation, so that time and fatigue may be saved and the animals maintained in a fit state. The harder the work the more the bulk of the food should be diminished and its nutritiousness increased; and to ensure this the hay should be reduced in quantity and the oats increased in proportion to the demands made on the physical energies; always remembering, however, that a certain amount of bulk is a physiological necessity, and the horse cannot live upon oats alone. Chopped hay and crushed oats dispense with an immense amount of mastication, while thorough assimilation is secured, waste averted, and strength and time are saved.

"The quantity and kind of food required by horses will depend, of course, upon the work demanded from them; insufficiency or inferior quality will not maintain vigour, while more than is necessary tends to plethora—a condition which has its risks to health, and also implies waste of forage. The hay and corn market is not so expensive as the horse market, and there is no economy in underfeeding. Bad food makes bad horses, and insufficient food produces weakly ones.

"Oats and hay should be sound and good. In judging of them for food it is well to remember the characters by which good may be differentiated from bad oats. In the first place, each grain consists of two parts, husk and kernel, the latter possessing considerable alimentary value, and the former scarcely any at all; so that oats which contain the largest proportion of kernel are those which are most serviceable to the horse. The relative proportions of kernel and husk vary considerably in different kinds of oats. In some samples the husk forms as much as 35 and 40 per cent., while in good grain it may be as low as 20 per cent. It is of importance sometimes to estimate quickly the feeding value of oats, and this can readily be done by separating the kernel from the husk by hand in a number of seeds, and then weighing each. This gives a better and a more practical indication than is afforded by the external appearance of the grains, their colour, or their weight collectively. It may be noted, besides, that oats which have the smallest proportion of husk are those which are most readily and thoroughly digested; and, as already mentioned, crushed oats are more quickly and perfectly digested than when they are whole.

"The weight of the oats is not altogether a trustworthy index to their nutritive value, though it is that which is generally adopted; the thickness of husk and its closeness to the kernel, as well as the dryness of the grain, will influence its density; so that there is often a rather wide diversity in different samples, with regard to their natural weight and nutritive value. The ordinary oats, which weigh only 38 lbs. to the bushel, are not very economical for feeding, and especially if they come from Sweden or Russia, where their quality—and particularly that of the Swedish oats—is rather low. It is better to give a smaller quantity of heavier thin-skinned oats.

"Of course, the oats should be sound, when musty they are likely to do great damage. And the same may be said of the hay. This varies considerably in feeding value, according to not only the grasses which enter into its composition, but also according to the situation, the soil, the district, and even to the countries in which it is grown; the manner in which it is preserved or made also influences its value as food.

"Other grains besides oats are sometimes substituted for these, wholly or in part. Maize is one of them, and is somewhat largely in use for omnibus and tramway horses. When the seeds are broken it is very digestible and economical, and may replace one-third, two-thirds, or even the entire ration on occasion; but good oats are preferable, as they sustain animals which are undergoing severe labour much better, and do not soften the liver like maize.

"Beans are a valuable adjunct to the food of hard-working horses when given in the proportion of one-tenth or one-twelfth to the other grain, and the same may be said of peas.

"With regard to the quantity of food required by horses, this, as has been stated, should depend, over and above a certain amount required to maintain health, upon the work exacted, the size of the animal, and also, to some extent, to the degree of appetite. Something will likewise depend upon the mixture of grains, in which the object is generally to furnish what is deficient in one kind of grain by adding another which contains it in larger proportion; a course which is advantageous from a dietetic, and often also from an economical point of view.

"I have elsewhere insisted that to maintain a just balance between food and work, which the condition of the horse will pretty accurately demonstrate, the owner must be ready to increase, and as promptly diminish, the grain allowance as demands upon it are created or disappear. If the quality of the food is not sufficient to furnish material for the repair of waste tissue, the deficiency must be met by the consumption of an increased quantity. But, as has been pointed out, an excessive supply of comparatively innutritious food to compensate for defective quality is not only embarrassing to the stomach, but hampers the horse with bulky dead weight. Severely worked horses require much more reparative material than those which are not so taxed, and they should therefore be supplied with more concentrated food, easier of digestion, and rich in flesh-forming qualities.

"The chief tramway company in London, for instance, gives: maize, 13 lbs.; oats, 3 lbs.; beans, 1 lb.; hay and straw in chaff—of the first 7 lbs., the second 3 lbs.; while the Edinburgh Tramway Company allows: oats, 8 lbs.; maize, 4 lbs.; beans, 4 lbs.; hay, 14 lbs.; Marshlam, 2 lbs. In Paris these horses received in 1886: oats, 5.50 lbs.; maize, 12.92 lbs.; beans, .10 lbs.; bran and carrots, .50 lbs.; hay, 8.62 lbs.; straw, 7.30 lbs.

"The scale of rations for our troop-horses is usually 10 lbs. of oats, 12 lbs. of hay, and 8 lbs. of straw per day, the latter being used for litter, and the hay is rarely chopped. When in camp 2 lbs to 4 lbs. extra of oats are allowed, but no straw.

"For hunters during the season the grain allowance is high, from 16 to 18 lbs., with 8 to 10 lbs. of hay, and 2 or 3 lbs of carrots per day. Frequently 1 or 2 lbs. of beans are added to the ration.

"Carriage-horses, when hard worked, should be fed like hunters; ponies and under-sized horses do not require so much hay or grain.

"When horses require to be fed during work, grain should be chiefly given, the bulky food being allowed at resting time —as at night. Care should be taken not to overfeed horses at any meal; and if the grain is not mixed with chopped hay, then it should be given alone, and the hay allowed afterwards.

"Horses ought not to be fed, if possible, soon before going to work, but sufficient time should be given for digestion to be well advanced before exertion is undergone. Food should be allowed during the day at intervals of three or four hours, and long fasts ought to be avoided, as well as hurried feeding.

"The former leads to imperfect mastication and over-distension of the stomach when food is offered, with consequent indigestion; and the latter has a like result. When a long fast is unavoidable, then a quantity of warm gruel or a little mash should be given, to be followed by the ordinary feed shortly afterwards.

"An important point in feeding is to apportion the feeds in such a way that each will be consumed at the time it is given.

"Care is necessary in allowing water to horses. It should never be given soon after feeding, but always before it, and especially if the food is grain. If the horse is very thirsty, the thirst may be assuaged and the feed given a short time afterwards; if any more water is needed it ought not to be offered within two or three hours after feeding. When horses can always have access to water, they drink less, and so run less risk of indigestion and colic than when it is only offered at wide intervals. It is, therefore, the best plan to allow them to have water, like their food, frequently; if properly watered, they will not drink any more than is necessary for them. While undergoing severe exertion, they should receive very little. There are circumstances when it may be necessary to restrict an unlimited supply of cold water, as when a horse is exhausted from fatigue, has undergone prolonged abstinence, or when very cold, or even excessively hot. In such cases, a small quantity only should be allowed until the body is in a fit state to receive more; though a larger quantity may be given if it is tepid, or in the form of gruel.

"It should be unnecessary to add that water given to horses ought to be clean and fresh."

The Care of Young Foals.

The following observations on the rearing of young foals are offered for the consideration of persons who, engaged in horse breeding, may yet not have acquired that practical experience of its risks and requirements which is essential to guard them from unnecessary trouble and loss. For upon the care bestowed on foals during the early months of their existence, will almost entirely depend their immunity from disease, and their subsequent vigorous growth and perfect development. To those who have had ample experience, directed by intelligent observation, the information I venture to give may be altogether superfluous, though I have more than once been consulted by such persons on some of the subjects to which I am about to briefly refer.

It is acknowledged by those who have had much to do with foal rearing that very much of its success depends upon the manner in which the mares are treated during pregnancy, and immediately before and after parturition. The food and the exercise they receive, or the work they may have to perform, are important factors in the business, as idleness and obesity are not conducive to the production of vigorous healthy foals, any more than is overwork, bad or insufficient food, or any other debilitating cause. If mares must be worked during pregnancy—and judicious labour is undoubtedly beneficial—then they must be liberally fed, in order that not only their own system may be maintained in good condition, but that of the fœtus may receive a due amount of nutriment. Grass alone will not suffice, and a certain allowance of oats is necessary, with hay in addition. Oats are the best grain for in-foal mares which require this addition to their food, and they should, if possible, be crushed. Maize is not to be recommended, as it is stated that when this grain constitutes a principal part of the ration, the foals always show weakness of joints and muscles. Even when mares are running out at grass it may be advisable to allow some hay, and even oats under certain conditions of weather or states of health. To have thriving progeny the mares themselves should be strong and lively during pregnancy and after parturition.

It is also recognised that the period when mares are to foal, and the management calculated to regulate that event, demand considerable attention. The best months for foaling are, doubtless, March, April, and May, the last especially, as then the young creatures are almost certain to have genial weather, and nothing in the shape of food is comparable with the green herbage of spring and early summer for milk production in the dams. Early foaling is only too frequently synonymous with debility, unthriftiness, and stunted growth in the foals unless artificial treatment is adopted, and even hay and oats do not fully compensate for the absence of grass as an article of food.

It is only too well known to breeders that when foals miss a good start at the commencement of their life, and sustain a check to their growth, it generally requires much time and nursing to repair the damage; indeed, sometimes the effect is so serious that their vigour and full development are permanently arrested. Foaling late in the year is also objectionable, as the young animals have then not sufficient time to gain strength before the advent of winter.

The season of the year and state of the weather will determine the propriety of turning the dam and foal into the paddock or pasture after parturition, but the sooner this can safely be done, the better for both, if only for an hour or two at first, while the weather is fine; as the genial rays of the sun have a most exhilarating influence on the foal. Exposure to rain must be rigorously avoided, as the woolly texture of the foal's coat retains the wet for a long time, and is very likely to give rise to catarrh or some bowel affection. Sometimes mares, and most frequently those with their first foal, do not secrete a sufficient quantity of milk to nourish their offspring. Gentle rubbing of the udder with new milk, and allowing the foal to go to the teat as often as it will, stimulates the gland; while soft succulent food, such as grass, sloppy mashes of boiled barley or oats to which treacle has been added, assists in exciting the secretion. When the mare chances to be ill or dies, or does not give milk, then the foal must be nursed by a foster mother, or fed artificially with milk obtained from a mare or she ass. If this cannot be conveniently procured, then cow's milk and water, in the proportion of two of the former to one of the latter, sweetened with a little sugar, answers in the majority of cases. In those instances in which this food does not prove suitable, less of it may be given, and a preparation of husked beans boiled to a pulp and squeezed through a hair sieve, when it forms a thick fluid like cream, has been recommended as an excellent substitute.

A dose of castor oil, to the amount of one or two ounces, may be required by the foal so fed, as constipation is not unfrequent; and, indeed, this should always be given when the young creature does not obtain the first milk of its dam, and also when it is being suckled by the mare if its bowels are torpid. It is always judicious to notice the state of its bowels, as these are constantly liable to derangement while the foal is being artificially fed or suckled—constipation or diarrhœa being the most common disorders. Constipation sometimes occurs in a day or two after birth, and unless attended to promptly may entail serious consequences in a short time. Regulating the diet of the mare, giving her frequent bran and linseed mashes and other sloppy food, often gets rid of this condition in the foal. If it does not, then a dose of castor oil and an enema, if the constipation is obstinate, will generally afford relief.

Diarrhœa is more often a source of trouble with foals than constipation, and is in many cases fatal in a comparatively short time. Its causes are more or less obscure, but the food of the mare, and bad sanitary arrangements, are generally blamed. The diet of the mare should be changed, and crushed barley given to the extent of one or two quarterns daily, with a diminished allowance of grass and an equivalent of good hay with fresh clean water; while cleanliness in the surroundings should be observed, or the mare and foal removed to another place. The foal ought to receive a dose of castor oil with a drachm of carbonate of soda, and ten to twenty drops of chlorodyne in a little tepid water. Half a drachm of the carbonate of soda and the chlorodyne may be afterwards given twice a day in rice gruel, made by boiling rice to a jelly. It may be necessary to withhold a portion of the mare's milk and give this rice gruel instead. The foal's body should be kept warm and dry, and the hind quarters and legs clean.

Sometimes mares give too much milk, and if the foal is allowed unlimited access to it soon after birth, its digestion may become deranged. As a matter of precaution, a portion of the milk should be drawn from the udder before the foal is permitted to suck, but this need not be continued for more than a few days.

The period of weaning will depend upon circumstances, such as the quantity and quality of the milk the mare yields, her constitution and condition, and whether she is again in foal. The age of the foal itself is also a matter for consideration; but under ordinary circumstances it is generally agreed that September is a good month in which to take the foal from the mare, though in this allowance must be made for foals which are born early or late. Weaning should be a gradual process, and should inflict no injury on dam or progeny. Foals begin to eat oats at a very early age, and they should be encouraged to do so very soon, especially when two or three months old. Crushed oats are preferable to those which are whole, and if these are scalded and mixed with a little bran and boiled linseed, and a small quantity of salt, all the better. The quantity of oats that should be given will, of course, vary with circumstances, but more will be required after weaning than before. After weaning, if the foal is robust it will consume about two quarterns of oats daily, and bran mashes twice or thrice a week are not to be neglected. Beans have also been highly recommended before and after weaning. One authority asserts that half a pint of beans, gradually increased to a quart per day, supplied before weaning, will be of greater benefit than triple the quantity allowed at two or three years old.

It is bad policy stinting young foals in their food, and a

liberal allowance of that which is nutritious and suitable for vigorous growth is always profitable. More especially is this the case during the first autumn and winter after weaning, when good feeding is absolutely necessary to enable the young animal to withstand the weather, and to compensate for the loss of the mother's milk.

It has been observed that worms sometimes annoy foals exceedingly when they have attained the age of three or four months, or even earlier, but more particularly when they are yearlings. An examination of the fæces will generally reveal the presence of these parasites, while the appearance of the young animals, their staring, harsh and unthrifty-looking coat, longer than it should be, large pendulous belly, loss of flesh, with frequently a dry, husky cough and constipation alternating with diarrhœa, betray the effects of the worms. The foals should have access to rock salt, and small doses of powdered sulphate of iron given morning and evening in a little mash. Ten to fifteen grains of calomel given in mash, and repeated after an interval of ten or twelve hours, is a very effectual remedy, from four to six ounces of linseed oil being administered six hours after the last dose.

Warm and comfortable shelter during cold and wet weather, and attention to feeding, will ward off many of the maladies to which young foals are otherwise liable, but there is one disease which, if all accounts are true, is on the increase, and is sometimes very destructive to foals soon after birth, but does not appear to be much influenced by the conditions in which the animals are placed. It manifests itself by high fever, intense inflammation of the joints, more especially those of the knees, stifles and hocks, running on to formation of abscess and ulceration of cartilage and bones. The pain and suffering cause rapid emaciation, debility and death. Little can be done in the way of curative treatment, but much may be accomplished in the way of prevention. The cause of the disease is the entrance of specific germs into the wound at the navel or end of the navel string, before this has completely healed up after birth, and to prevent the admission of these dangerous organisms the greatest cleanliness is necessary, not only of the wound itself, but of the stable or shed in which the mare and foal are kept. If a number of foals are reared in the same establishment, the appearance of the disease should be the signal for immediate attention to the others. This should consist of daily dressing of the navel cord or sore with some disinfectant, such as carbolic acid and olive oil, one part of the former to fifteen of the latter, applied with a bit of sponge; or after the wound has been cleaned with tepid water, the part should be well covered either with powdered boracic acid, or equal parts of iodoform and starch powder, and covered with a piece of carbolised lint or fine tow, maintained in its place by a wide cotton bandage round the body. In about a week there will be no more danger. This treatment should be resorted to soon after birth.

Whether young foals are reared in straw yards or at pasture, or both, the hoofs require attention, and more especially in straw yards, where they are inclined to grow long and irregular in shape, which again is apt to react upon the limbs and cause their deviation from a good direction. A little judicious management here may save much trouble and disappointment afterwards.

When foals run about on very hard ground, not only are the hoofs sometimes too much worn and the feet consequently tender, but the concussion may injure the bones and joints of the limbs, and it is probable that some of the diseases of these which are supposed to be hereditary may be originated in this way in early life. "Cecil," many years ago, drew attention to the damage sometimes done to the hoofs from hard dry ground, and recommended that a couple of barrowfuls of clay or soil retentive of moisture should be deposited in a part of the yard or paddock where the manger or receptacle for food is placed, so that the foal might stand in it during the time of feeding; this soil is to be kept soft with water when moisture is required, and a little common salt may be occasionally sprinkled on it with good effect. In many cases the clay may be dispensed with, by merely throwing water on the spot where horses stand to feed—that is, unless the soil is very sandy and dry.

The desirability of accustoming foals at an early age to have their legs and feet handled must be evident, and in practising them to this manipulation progress will have been made in teaching them to allow their hoofs to be trimmed and regulated by means of the knife, or, better, the rasp.

"GOOD AT TIMBER."

CHAPTER XI.

DISEASES AND INJURIES TO WHICH LIGHT HORSES ARE LIABLE.

In the following remarks on the diseases and injuries to which the lighter breeds of horses are liable in a state of domestication, it is not intended to give such information as will enable the horse-owner to play the part of veterinary surgeon, and treat his animals in every case as if he were a person thoroughly trained in veterinary medicine and surgery. Such endeavour would be as futile as it would be inexpedient and dangerous. Printed directions and horse-doctor books cannot do this; the utmost service they can yield is to afford the attendant upon, or the owner of, horses some idea of the disorders and accidents to which these creatures are exposed, so that he may be able to form an idea as to what should be done before the arrival of the veterinary surgeon, in cases of emergency, or when the assistance of this useful individual cannot be readily obtained. The majority of horsemen now-a-days have received some kind of instruction in horse-management, either at one of the several agricultural schools established in the United Kingdom, or by attending the lectures and demonstrations so frequently given in various parts of the country, through the instrumentality of agricultural societies or County Councils. In any case, for the treatment of the more serious diseases and accidents, far more experience and skill are needed than are possessed by the amateur, however well read he may be in veterinary books; so that, in order to avert loss and damage, it is the wisest course to invoke professional aid without delay, resorting to such measures as may be deemed appropriate until its arrival.

In the limited space at our disposal, only a few of the more frequent diseases and accidents can be referred to, and these briefly.

Fever.

Symptoms.—Fever is a condition of the body in which the temperature is higher than in health. The ordinary temperature of the horse's body—what is termed the internal temperature—is about 100° Fahrenheit. It is best ascertained by the self-registering thermometer, which is inserted into the rectum and kept there for a minute or so. When this temperature rises above 101°, fever is present; if it reaches 104°, then the fever is somewhat serious, and when it gets to 106° it is very severe. In proportion to its height the horse becomes wasted and debilitated.

The pulse, which is usually 38 or 40 beats a minute, and is best felt on the inner side of the lower jaw, is correspondingly increased, and the beats may reach 60, 80, or even 100 per minute, though when it is over 80 the fever may be said to be high. The breathing is also quickened, the number of respirations—which are about 8 per minute in health—increasing in a corresponding manner with the pulse. Coincidently with these phenomena the skin is dry and hot, though exceptionally it may be wet with perspiration; the mouth is also dry, hot and pasty when the finger is passed into it, and it generally has the odour of indigestion. The appetite is either much diminished or lost, and though the horse may drink a good deal of water, the urine may be less in quantity and high coloured. Sometimes the breath feels very warm, and the eyelids are swollen, with perhaps tears running down the face; in certain cases the horse is somewhat excited, in others he is listless, apathetic and depressed.

Fevers are of several kinds—such as continuous, remittent and intermittent, according to their course; and simple, specific, inflammatory, adynamic or hectic, according to their symptoms and cause.

In many cases, at the very commencement of fever there are signs of rigor or chill, the coat being then lustreless and hair erect, and the skin cold wholly or in parts; while the horse may even be trembling slightly. The diminished desire for or refusal of food is always a very significant sign of commencing illness in a horse, and should therefore receive immediate attention.

Treatment. — The causes of fever are numerous, and its successful treatment largely depends upon the cause being ascertained. This is discovered by noting the symptoms and inquiring into the history of the case. This needs tact and skill, and as some of the fevers are very serious and soon run on to a fatal termination, it is advisable to obtain veterinary advice in good time. The amateur, however, can assist in the treatment by having the horse moved into a well ventilated horse-box or stable and made comfortable, but not oppressed, by means of clothing on the body and bandages to the legs; if the latter and the ears are cold—which is sometimes the case—then they should be hand-rubbed. The horse ought to be allowed plenty of cold or tepid water to drink, with sloppy food. Nursing is the chief means by which restoration to health can be secured. Medicines must be sparingly given by the unskilled, and at most nothing more should be administered than about an ounce of nitrate or carbonate of potass in a bucket of water once or twice a day. If he will lie down, the horse should have a good soft bed. He ought not to be exercised until the appetite has returned, nor put to work until he feels well, and has regained his usual strength and spirits.

As nearly all young horses brought up from grass or from the country to town stables are liable to an attack of town or stable fever, they should be put into airy stables, and carefully fed and exercised until they have become seasoned somewhat. And when put to work, this should be light and only for a short time at first.

Catarrh.

Catarrh, or what is termed a "cold in the head," may attack old and young horses alike, and at any season of the year; though it is most frequent in cold or changeable weather. One of the great predisposing causes is a hot and badly-ventilated stable.

Symptoms.—There is more or less fever at first, with sneezing, perhaps shivering, cold legs and listlessness, and slight loss of appetite. Soon there is a discharge of watery fluid from the nostrils, sometimes also from the eyes; this quickly becomes yellow and purulent, and not unfrequently cough ensues, with sore throat and more or less difficulty in swallowing. Very often, too, these symptoms are accompanied by more fever, loss of appetite, and swollen glands about the upper part of the throat.

Treatment.—The treatment chiefly lies in nursing, making the horse comfortable by body clothing and leg bandages, keeping the stable at a moderate temperature and well ventilated, and giving mashes of bran and linseed, with small quantities of nitrate of potass in the drinking water. The head may be held over a bucket of boiling water in which there is some hay and a little oil of turpentine or carbolic acid, so that the steam may pass up into the nostrils. If the cough is troublesome, the upper part of the throat may be well rubbed with soap liniment, or a liniment composed of equal parts of olive oil, oil of turpentine and spirit of hartshorn. Should the cough be very severe, a little tincture of opium or chloroform may be dropped in the bucket of hot water, and a sack or blanket thrown over it and the horse's head in order to keep in the vapour.

Should the horse be debilitated after the more severe symptoms have disappeared, a drachm of powdered sulphate of iron may be mixed in the mash once a day.

Strangles.

This is an infectious disease to which young horses are more especially predisposed, and somewhat resembles catarrh. One attack generally protects horses against a second. There is great probability that every case of strangles is due to infection, and from this point of view, and also because of the trouble and damage it only too often occasions, it should be treated as a communicable disease, so as to prevent its spreading.

Symptoms.—It generally commences with fever and dulness, and disinclination to eat. The throat generally begins to feel sore, and there is difficulty in swallowing, while the glands between the jaws and below the ear are swollen and painful to the touch. In nearly all cases there is inflammation of the air-passages of the head, and this is manifested by a discharge of yellowish matter from the nostrils; there may also be cough. The swelling between the jaws increases in extent and painfulness, and not unfrequently this causes obstruction to the breathing, which is marked by a noise both in inspiration and expiration. In some cases this obstruction is so great that suffocation is imminent, and to prevent it the windpipe has to be opened lower down the neck, and a tube inserted through which the horse can breathe. In the usual course of the disease an abscess forms in the middle of the swelling, and when this bursts the animal is generally relieved; the swelling subsides, fever rapidly diminishes, swallowing becomes easier and the appetite is increased.

This is the ordinary course of the disease, but sometimes it runs an irregular course. The fever persists, and the other symptoms may increase in intensity; swellings appear in different parts, and may form abscesses, or disappear and reappear elsewhere, and the disease may continue for a very long time; in the simple form it seldom lasts longer than a fortnight or three weeks, whereas in this malignant or irregular form it may run on for one or two months, or even longer. In some instances it is a year or two before the animal completely regains a healthy and robust condition. This protracted phase of the disease is due to the repeated occurrence of abscesses in various parts of the body; these suppurate, heal up, and are succeeded by others; they sometimes form in the internal organs, and then usually cause death.

A not unfrequent sequel of strangles is "roaring," which greatly depreciates the animal's value, as it interferes with the breathing.

Prevention.—Strangles should be dealt with as a very contagious disease, and careful isolation of those affected, with disinfection measures, ought to be strictly observed.

Treatment.—Good nursing must form the chief part of the treatment of strangles. Whenever a young horse shows signs of ailing, it should be placed in a well ventilated and moderately warm stable or loose box—the latter is always the better; this should be kept clean and comfortable. If the weather is cold a blanket may be worn over the body, and it may even be necessary to place woollen bandages on the legs if they have a tendency to become cold. The food should be soft, and consist of bran and linseed mashes, oatmeal gruel, and a little good meadow hay, with now and again some scalded oats. If in season, grass and carrots or sliced turnips are good. The water given to drink should have the chill taken off if the weather be cold, or the oatmeal gruel may suffice. A little nitrate of potass—say half an ounce—may be put in the drink now and again.

If the fever runs high, a fever draught may be given (this is also useful in ordinary fever and catarrh). A useful draught is composed of acetate of ammonia in solution, three or four fluid ounces; sweet spirits of nitre, one ounce; bicarbonate of potass, half an ounce; to be mixed in a pint of tepid water. This draught may be given once a day until the fever abates. Should the breathing become noisy, or the horse experience much difficulty in swallowing, then hot water vapour should be inhaled as in the treatment of catarrh, a little carbolic acid or oil of turpentine being added to the water. At the same time, the white liniment recommended for sore throat in catarrh should be applied to the upper part of the throat and beneath the jaws where the swelling takes place. Sometimes, when the swelling is very extensive and dense, it is well to apply a hot linseed meal and bran poultice to it, or to blister it with cantharides ointment.

The abscess may be opened when it is fully formed, which is ascertained by its "pointing" and feeling very soft at a certain part, or left to open spontaneously, which is the better course unless the amateur is sufficiently skilled in using a lancet. When it is opened the wound should be kept very clean by washing with warm water and a sponge, and dressed with a solution of carbolic acid—one to fifty of water.

If the fever has been high or the abscesses large, there is often a good deal of debility supervening, and this must be combated by a generous diet, such as scalded oats and boiled linseed to which some salt has been added. If there is very much prostration and the digestion is impaired, it may be necessary to give a pint of milk two or three times a day; to this a teaspoonful of carbonate of soda should be added. Sometimes it has been found advantageous to give one or two eggs beaten up in milk in the course of the day, or a pint or quart of stout or porter, morning and evening.

In the irregular form of strangles the same system of nursing should be carried out, and the abscesses opened wherever they appear. Sulphite or salicylate of sodium may be given in half-ounce doses in water twice a day.

A stable or loose box which has been occupied by a horse affected with strangles should not again be used until it has been thoroughly cleansed and disinfected.

Influenza.

This is undoubtedly an infectious fever, which appears in a very extensive manner over large tracts of country, the outbreaks always occurring where there is much movement of horses from one place to another; in this way it follows the lines of traffic, and may appear at any season of the year. Horses of all ages and under all kinds of conditions may be affected, but it generally visits most severely those which are badly attended to and kept in unhealthy stables.

Symptoms.—The most marked characteristic of influenza is the intense prostration that accompanies the fever from the very commencement; otherwise, in most of the outbreaks the symptoms are much the same as those of catarrh, and they may all be developed very quickly. Sometimes the air-passages and lungs are chiefly implicated; at other times the abdominal organs suffer most; and in some of the outbreaks symptoms of rheumatism, with swelling of the legs, head and other parts of the body, predominate. Not unfrequently we may have all these symptoms manifested in one animal. The disease has received several names according to the prevailing symptoms. The catarrhal symptoms may be well marked, and then we have, in addition to the fever and great debility, the signs of ordinary catarrh; these, under favourable conditions, gradually subside in eight or ten days, and in a fortnight or three weeks the animal has usually recovered.

When the lungs and bowels are implicated, however, the cases are more serious, especially if the sanitary conditions are bad and the horses are not healthy and vigorous.

Treatment.—One of the essential conditions in the successful treatment of influenza is relieving the animal from fatigue and work whenever the first signs of illness become apparent. These signs are generally diminished appetite, listlessness, weakness, dry hot mouth, hanging head, swollen eyes, and perhaps shivering. To work and fatigue the horse after the disease has seized him, is to expose him to the risk of a more severe attack than he otherwise would have, and may lead to his death.

Therefore cessation of work at once is all-important. Good nursing comes next in importance, for the amateur—and even the veterinary surgeon, for that matter—can do little more than place the patient in the best possible hygienic conditions and maintain the strength. More horses are injured than benefited by the administration of drugs in this and many other diseases. Good ventilation, keeping the horse's body warm and comfortable, and giving soft and easily digested food, are the chief points to be attended to. If the symptoms are mainly those of catarrh, then the treatment should be the same; if the chest is affected, the treatment should be the same as for pleurisy or inflammation of the lungs; and when the bowels are implicated the treatment prescribed for inflammation of them must be adopted. When the legs and other parts of the body swell, then they should be kept as warm as possible by means of woollen bandages and rugs. Salicylic acid should be given in one-drachm doses in a little thick gruel twice a day. When the animal is recovering, in order to counteract the debility, it is advisable to give vegetable and mineral tonics. The best of these for the horse are powdered gentian and sulphate of iron—an ounce of the first and two drachms of the second in a ball once or twice a day. Boiled linseed is advantageous.

The horse should not be put to work until quite recovered, and even then this should be rather light for some time.

Glanders and Farcy.

These are not two diseases, but only one disease in two forms; we shall therefore treat of these as one disorder under the name of glanders.

Glanders is a virulent disease special to horses, but transferable from them to several other species of animals and to mankind. It may affect every part of the body, but is most frequently witnessed in the head or on the skin. It may be chronic or acute, but it is generally the former; both are marked by fever, which is most severe in acute glanders. It is very contagious, and can be produced by giving the poison in the food or water, or in a ball; and it can gain introduction to the system by inoculation, through a wound or abrasion, and in other ways. Contact with glandered horses, being put into stables which have been inhabited by them, drinking out of water-troughs they have frequented, or eating from receptacles they have fed in, are the usual ways in which healthy horses acquire the disease. It is most frequently witnessed among large studs of horses, and especially those which are over-worked, improperly fed, or badly housed. A variable period elapses between an animal receiving the poison and the appearance of the first symptoms, but it is between a week and several months. The poison is contained in the discharge from the nostrils and from the sores, as well as in the blood and other fluids; but the disease is mainly spread by means of the matter from the nostrils and sores. In the ass and mule glanders nearly always appears in the acute form and rapidly runs its course.

Symptoms.—The symptoms in acute glanders are much more marked than in the chronic form, but the high fever constitutes the chief difference. This fever lasts for a few days generally, then subsides, but only to reappear after a short interval. There is much depression, and the animal does not care to move. There is usually a discharge of a yellowish sticky matter from one or both nostrils, which adheres around them, and at the same time there is one or more sores or ulcers inside the nostril on the partition separating the nostrils. If the discharge is only from one nostril, then the sores are on that side. When the ulcers are deep, the discharge may be streaked with blood. The glands inside the lower jaw are also enlarged, hard and knotty. Ulcers may or may not appear on the skin at the same time. Sometimes the ulcers are high up in the nostril and cannot be seen, and not unfrequently they extend down the windpipe. The lungs are generally implicated, or they may be alone the seat of disease, but this is more frequently the case in chronic glanders. In the acute form, if the horse is not killed it dies from suffocation or exhaustion.

The chronic form only differs from the acute by the severity of the symptoms. A horse may live for a considerable time when affected with chronic glanders, and even perform hard work, as the constitutional symptoms are comparatively slight. But this form always terminates in acute glanders if the horse is not destroyed.

Farcy is merely superficial or skin glanders, and it also may be acute or chronic. There are ulcers on various parts of the body, and these generally discharge; they are connected by a prominent line or cord. The legs are most frequen.ly involved, and then they are generally swollen and painful, an the horse moves with difficulty. Farcy generally terminates in glanders.

Treatment.—Glanders is practically incurable, and owing to its dangerous character its cure should not be attempted. Diseased horses should be at once destroyed, and those with which they have been in contact, or which have stood in the same stable with them, ought to be considered suspected and consequently kept apart from the others. Stalls and places which have been occupied by diseased and suspected horses, should be thoroughly cleansed and disinfected.

Bronchitis.

Bronchitis is inflammation lining the windpipe and its branches in the lungs, and is usually due to colds, though it is sometimes a complication of other diseases, and it may even be produced by the entrance into the air-passages of irritant fluids or gases.

Symptoms.—Bronchitis may be acute or chronic, but in young horses it is most frequently the former. This generally begins with shivering and dulness; then fever sets in, and the breathing is quickened, while there is a hard loud cough. There may or may not be a discharge from the nostrils at first, but there is generally after a day or two, and in a few days it may be quite copious. The cough increases in frequency and severity, and is very exhausting, while the appetite is much diminished. Death may ensue from filling up of the bronchial tubes with matter. But a favourable result may be anticipated when the fever gradually subsides, the cough becomes softer and less frequent, and the discharge from the nostrils less and thinner in consistency.

Chronic bronchitis is generally seen in old horses. There is little, if any fever, and the nasal discharge is very trifling, the most marked symptom being the cough, which is often very harassing.

Treatment.—As bronchitis commonly occurs in cold weather, the horse should, if possible, be put into a comfortable well-ventilated stable or loose box, and the body clothed, the legs being enveloped in woollen bandages or straw bands, after being well hand-rubbed. Hot water vapour, into which a small quantity of oil of turpentine or carbolic acid should be put, ought to be inhaled by the animal, as for catarrh; and the throat should be rubbed with the white liniment already mentioned, or with compound camphor liniment. The same liniment may also be applied to the sides of the chest, or this may be enveloped in a thick blanket, and hot water (not scalding) poured on it for an hour or two at a time; the blanket must then be removed, the skin thoroughly dried, the liniment rubbed in, and a dry blanket put on. A draught composed of one drachm of camphor, two ounces of solution of acetate of ammonia and an ounce of nitric ether, mixed up in about ten ounces of water, should be administered twice or three times a day. The diet should consist of mashes of linseed and bran, with a few scalded oats; carrots or green food should also be allowed, and a little good hay. When convalescence is setting in, a drachm of powdered sulphate of iron may be given in the mash twice a day, and the food may be more nutritious.

Little can be done for chronic bronchitis beyond keeping the horse in a cool, well ventilated stable, clothing the body comfortably, giving easily digested food, and allowing steady slow work.

Congestion of the Lungs.

No animal is so liable to congestion of the lungs as the horse, and it may be an accompaniment or sequel of the other diseases, or occur by itself. It usually appears in the acute form in the latter case, and it is this which will now be noticed.

Acute congestion of the lungs may be induced by sudden severe exertion when an animal is not in good condition, or by long continued severe exertion when in good training; it may also be caused by exposure to cold, and especially to cold winds and wet.

Symptoms.—The symptoms of acute congestion of the lungs are of a very marked character. The breathing is extremely hurried and laboured, the nostrils widely dilated, head carried low, countenance anxious and haggard, body usually covered with perspiration, legs stretched out and cold, the flanks heaving tumultuously, and sometimes the heart can be heard beating violently. Not unfrequently blood flows from the nostrils, and if this is foamy it shows that it comes from the lungs. If not quickly relieved the horse will die from suffocation.

Treatment.—This, to be effective, must be prompt. The horse should not be moved or disturbed, and if wearing harness this ought to be taken off. An abundance of fresh air must be allowed; the legs and body should be well rubbed and clothed, and if any turpentine liniment is at hand this should be applied to the legs before they are bandaged. Brandy or whisky in six-ounce doses may be given in water every hour or two hours for the first three doses, and then every four hours for four or five doses. If there is thirst, cold water, or, better, oatmeal gruel can be given. If the symptoms do not soon subside, hot water should be applied to the sides in the manner already indicated, and care should be taken to keep the animal from draughts of air.

After recovery, some days' rest should be allowed, and careful feeding observed.

Inflammation of the Lungs.

Inflammation of the lungs may be a disease of itself, or follow catarrh, bronchitis, congestion of the lungs, or other disorder, as well as be due to sudden chill, foul hot air in stables, &c; pleurisy is often present.

Symptoms.—There is fever, the pulse and respiration are increased, the animal is dull and dejected and wanders about in the loose box, but rarely lies down. There is frequently a short dry cough, and there may also be a slight discharge from the nostrils of rust-coloured mucus when the disease is advanced; the skin of the body and legs is cold, the mouth is hot and dry, and the membrane lining the eyelids and nostrils is deep red in colour.

Treatment.—This is similar to that for congestion of the lungs. Fresh air is above all things necessary; at the same time the body and legs must be kept warm. From four to six quarts of blood abstracted from the jugular vein sometimes lead to a favourable change in the case of fat, high-conditioned horses. At first the following draught may be given every four hours:—Fleming's tincture of aconite, six minims; nitric ether, one ounce; solution of acetate of ammonia, four ounces. To be given in a quart of thin gruel or tepid water.

If there is much debility, then instead of this draught six ounces of brandy or whisky may be administered three or four times a day in the same manner.

The food should be sloppy mashes of bran or linseed, with oatmeal gruel, a little good hay, and green forage or carrots; cold or tepid water may be allowed to drink, and in a bucketful of it an ounce of nitre may be dissolved.

When the horse is recovering, a drachm of powdered sulphate of iron may be given in the mash twice a day.

Pleurisy.

This is inflammation of the membrane lining the chest and covering the lungs, and may be a complication of pneumonia or other diseases, or exist independently.

Symptoms. — There is fever succeeding a shivering fit. There is most acute pain on moving the ribs, which causes the horse to keep them fixed as much as possible, and to breathe quickly in a careful manner with the abdominal muscles. The countenance looks distressed, and there is a short interrupted cough, while in attempting to turn there is heard a painful grunt. Pressure between the ribs causes acute pain. The horse does not lie down. Effusion into the chest very often sets in early, and then there is less pain, but the breathing becomes deeper and laboured, owing to the pressure on the lungs.

Treatment.—This does not differ much from that adopted in inflammation of the lungs. The general management should be the same, and the hot water applications of the chest should be even longer continued. Mustard may be applied to the sides of the chest with advantage. Nitrate of potass in ounce doses should be given in the water or gruel, and Fleming's tincture of aconite, in four or six-minim doses, given in a small quantity of water every three or four hours.

After three or four days, whisky in four-ounce doses may be given twice or three times a day in gruel. If fluid accumulates in the chest, then it should be gradually removed by a surgical operation, which the amateur had better not undertake.

Rheumatism.

Some horses are particularly liable to rheumatism, an inflammatory condition of certain structures in connection with joints, tendons, muscles, &c.

Symptoms.—Rheumatism may be acute or chronic. The acute form is accompanied with fever, and usually manifests itself suddenly in the joints of the limbs—as the stifle, fetlock, hock, knee, or sheaths of the tendons. There is great lameness and pain on pressure, and often more than one part is affected; not unfrequently the swelling and pain leave the joint as suddenly as they came, and attack another part. The heart is often involved. In bad cases the joints are much enlarged.

Treatment.—Hot fomentations to the inflamed parts, of water in which poppy heads have been steeped; with gruel in which ounce doses of the bicarbonate of potass have been dissolved. The animal should be kept comfortable, and if there is constipation a mild dose of physic may be given. If the fever runs high, salicylate of sodium in two-drachm doses three times a day should be given in a pint of water or gruel. When the inflammation in the joints or sheaths of the tendons becomes chronic, then it may be necessary to rub them with the white or soap liniment, or with a liniment prepared as follows:—Coutt's acetic acid, two ounces; whisky, two ounces; oil of turpentine, two ounces. One white of egg to be beaten up with these. The skin should be first well brushed, then the liniment should be firmly rubbed in.

Laminitis.

Heavy horses are more liable to inflammation of the feet, perhaps, than light ones; and the fore feet are much oftener affected than the hind ones. Many causes will give rise to it, such as bad shoeing, injuries, severe travelling in hot weather, indigestion, superpurgation, &c., while it is often a sequel of pneumonia, influenza, &c.

Symptoms.—This is a most painful disease, and is accompanied by a considerable amount of fever. The horse perspires, breathes quickly, and looks as if suffering intense agony; the symptoms might be mistaken for inflammation of the lungs, but attempting to make the horse move reveals the nature of the disease. He will not stir if he can avoid it, but remains rooted to the ground, resting his weight as much as possible on the heels. The feet feel extremely hot, and striking the hoof intensifies the pain.

Treatment.—The shoes should be removed from the inflamed feet, if possible, and the walls lowered to a level with the soles, so as to allow these and the frogs to sustain a greater portion of the weight. But this is a difficult operation, as the horse suffers excruciating pain when one fore foot is lifted. The animal should therefore be put into a sling, or, better, thrown down, the litter being peat moss or sawdust. This allows the shoes to be taken off and the feet attended to. Cold poultices of bran or other material, or cold wet cloths, should be applied to the feet and kept constantly wet and cold. Carbonate of soda may be mixed with the poultices or water. Unless there has been purging, a dose of physic should be given, and the diet ought to be of a laxative nature. If the horse is lying and does not attempt to change position, he should be turned over every day to prevent the occurrence of sores on salient parts of the body. When the intense pain and inflammation have subsided, exercise on soft ground should be enforced for some time.

Colic.

Colic is spasm of the intestines, or may be due to distension of these with gas (flatulent colic). Many causes may give rise to colic, such as indigestion, mismanagement in feeding or watering, chills, worms, &c.

Symptoms.—The attack is usually sudden, and the chief sign is the manifestation of restlessness, owing to the pain experienced. The horse lies down and rolls about, then gets up, shakes himself, looks towards his flanks, paws, strikes at his belly with the hind feet, and if in a loose box wanders around it. The pain subsides, and the horse then remains quiet and may commence to eat; but in a short time the symptoms reappear, and at each recurrence they may increase in intensity, and attempts may be made to stale, while the animal may perspire freely and manifest anxiety. In flatulent colic the symptoms are analogous to those in spasmodic colic, the chief difference being that in the former there is distension of the belly, and the breathing is therefore more interfered with; the horse also lies down more carefully and does not roll so much.

Treatment.—No time should be lost in treating cases of colic, and the relief of pain is one of the first objects to be obtained. Six ounces of whisky should be given in a quart of tepid water, and if two ounces of laudanum can be added to this so much the better. The belly should also be well rubbed with straw wisps. If there is constipation a dose of physic ought to be given; and when there is distension of the abdomen, after the stimulant just mentioned an ounce of oil of turpentine in a pint of linseed oil ought to be administered. The alcohol and laudanum may be repeated in three or four hours if the symptoms do not abate. An enema of soap and water every two hours is very serviceable in obstinate cases; and when the attack is very acute, blankets wrung out of very hot water and applied to the abdomen often act very beneficially.

Inflammation of the Bowels.

Like colic, which it often succeeds, inflammation of the bowels arises from many causes.

Symptoms.—These are not unlike those of colic, except that there is no intermission in the pain, which is much more severe, and the breathing and pulse are quickened throughout; the pain is also increased by pressure on the abdomen. The body is covered by profuse perspiration, and the expression is haggard and distressed. In this disease no alcohol should be given, nor yet laudanum, but, instead, powdered opium in two or three-drachm doses, rubbed up in flour gruel, every two or three hours; to this may be added twenty drops of tincture of aconite, two drachms of chloroform, or two ounces of sulphuric ether. Hot water should be applied to the abdomen by means of rugs, and the white liniment or mustard plaster may also be applied to this region before the hot water is resorted to.

When the horse can eat, the diet should consist of linseed and bran mashes, and no hay or other solid food should be allowed for some days.

Worms.

Worms are often troublesome to horses, and cause irritation of the intestines, and unthriftness and debility. There are several kinds of worms which we need not, for lack of space, describe, especially as the treatment is nearly the same for all. This generally consists in the administration of a purgative, followed by an ounce dose of oil of turpentine in flour gruel or well mixed in a pint of milk; or one or two one-drachm doses of tartar emetic in a little mash, followed by half-a-dozen one-drachm doses of powdered sulphate of iron—one dose morning and evening.

Lamenesses.

The horse is, from the nature of his work, much exposed to lameness, and this very often becomes permanent, and more or less reduces his value. Lameness may be due to many causes, and these may be in operation in any part of the limb or limbs; sometimes injury or disease of other parts of the body will also produce lameness. We will notice some of the more common forms of lameness, with their causes and treatment.

Sprains.

Sprains may occur to tendons and ligaments, less frequently to muscles, and this injury may be more or less severe and cause a proportionate degree of lameness. Ligaments and tendons, as well as muscles, during violent efforts or from twists, may be over-stretched and their fibres torn, or the injury to them may be brought about gradually, as in some tendons and ligaments of the lower part of the limbs. No matter where sprains occur, more or less prolonged rest, as complete as possible, is essential to rapid and permanent recovery. Next to rest comes reparative treatment, and this will vary somewhat according to the seat and nature of the sprain. When it is quite recent, attempts must be made to check the swelling and inflammation that ensue, and with this object in view the application of water—cold or hot—or soothing and evaporating lotions, is resorted to. All are beneficial according to the assiduity with which they are applied. The water should always be rather cold, or as hot as the horse can bear it. When it can be done, the part should be enveloped in bandages or swabs, so as to retain and distribute the moisture or lotion. Perhaps the best lotion is that composed of Goulard's extract (subacetate of lead) and spirit in equal parts, with eight or ten parts of water. When the pain and swelling have subsided somewhat, then a mild stimulant may be applied—such as acetic acid liniment already alluded to. Gentle exercise may also be allowed if there is no lameness, and continued until the horse is fit for work.

Sprain of the Back Tendons.

This is perhaps the most frequent sprain to which light horses are liable, and may occur either in the fore or hind legs. There is swelling, heat, and pain on pressure of the injured part, and lameness corresponding to the extent of the injury. A shoe raised two or three inches at the heels should be put on the foot of the sprained leg, and the general treatment prescribed above resorted to. If the injury is very severe and considerable thickening remain, it may be advisable to apply the biniodide of mercury ointment or cantharides ointment to it; it may even be necessary to "fire" the part in order to effect efficient recovery. Instead of this, the projection of cold water from a hose for from ten to twenty minutes three or four times a day, may be advantageous in expediting a cure; indeed, this may be carried out from the very commencement, the lead lotion being applied in the intervals. Sprains of these or other tendons or ligaments in this region may also be treated after the method recommended by Captain Hayes, which consists in enveloping the part in cotton wool, and bandaging tightly, in such a manner as to ensure uniform pressure. This bandaging may be employed after applying the hot or cold water or lotion, and is most conveniently carried out as he directs:—Take about half-a-pound of cotton wool, and a cotton bandage (such as can be got in any chemist's shop) about three inches broad and six yards long. First of all, wrap loosely round the leg a piece of soft cotton cloth, or put on an ordinary flannel bandage, as the contact of wool sometimes causes irritation to the skin. Place a little cotton wool at each side of the leg at the place where it is desired to commence, and loosely wrap the bandage over it, adding at each turn more cotton wool, some of which should also be placed at the front and back of the leg until there is a layer about four inches thick round the part. As the bandage is passed round the leg it may be gradually tightened, until at last it is made very tight, when it can then be secured by sewing or by tapes. The bandages should be removed after twenty-four hours, the part rubbed firmly upwards by the hand (the leg being held up during this massage, and flexed and extended), and a fresh bandage of the same kind put on. The bandage may then be removed morning and evening, and the part hand-rubbed, and passively worked by bending the joints without causing the horse to move. The tendon may be rubbed with stimulating liniment during the massage; if the hair is long it may be clipped off. The cotton wool should be of the ordinary kind, soft and elastic, and it is better to have it fresh at each application. The diet should be rather laxative, and green forage be given if it can be procured.

The high-heeled shoe should not be kept on the foot for more than a fortnight, when its heels may be gradually lowered. If considerable improvement has not taken place in three weeks of this treatment, a charge may be applied to the tendon. This is variously composed, but the usual ingredients are Burgundy pitch and bees-wax, four parts of each; when these are melted in an iron ladle, two parts of mercurial ointment are stirred in. When moderately warm this is plastered in a thick layer over the leg by means of a spatula or hard brush, pieces of cotton wool being stuck on the skin and the hollows on each side of the tendon as the smearing goes on. Over these the mixture is to be daubed, and when, finally, sufficient has been applied to make the leg a rounded mass, a long cotton bandage is tightly bound over it, the mixture being laid upon this at every turn, and cotton wool placed between each layer, so as to effect equable and firm pressure. If at any time the layers should become loose, they may be plastered with the warmed mixture. From three to five weeks is sufficiently long to keep on this bandage.

Splints.

Splints are bony tumours which form either inside or outside the leg, usually the former, and generally in the neighbourhood of the small splint bones. They most frequently form in young horses, and are readily seen when the limb is looked at in front. It is usually when they are forming that they cause lameness, but when they are so situated as to interfere with movement, the lameness may be permanent. There is heat and pain on manipulation.

The best treatment for the amateur to adopt consists in the application of Goulard's lotion, already described; this being poured on to a woollen or cotton bandage, enveloping the leg where the splint is forming. After a few days of this treatment a little piece of the biniodide of mercury may be rubbed into the skin over the tumour. Exercise should be allowed on soft ground.

Ring Bone.

This is a deposit of bony matter on the surface, front or sides, of the pastern bones, and is generally very serious, owing to the deposit interfering with the tendons and ligaments covering it. It is most frequently observed in the front pasterns.

The treatment should be the same as for splints, but it must be long continued, and the horse should be rested as much as possible, the stall or loose box being laid with peat-moss litter. In chronic cases firing may be necessary.

Side Bone.

This is comparatively rare in light horses, and in them it is limited to the coarser breeds. The name is given to the lateral elastic cartilage on each side of the foot, towards the heels, when it becomes hard and rigid from deposition of bony matter in its substance. This may occur from injury, such as a tread, but more frequently it is due to inflammation set up in the cartilage from some internal cause. Bad shoeing may give rise to it, as when one side of the hoof is left higher than the other, and so causes twisting of the limb. Sometimes there is lameness, and there is nearly always impaired action.

Treatment.—Little can be done in the way of treatment. Repeated blistering with the biniodide of mercury ointment, exercising on soft ground, keeping the hoof level, and allowing the frog to come in contact with the ground, or shoeing with a bar shoe, are the measures likely to render the horse so affected serviceable.

Navicular Disease.

Navicular disease is almost entirely confined to light horses, and is the most serious, and also one of the most frequent causes of all lamenesses. There may be a hereditary tendency to it in the way of predisposition or conformation, but it is rarely seen in young horses, and most frequently attacks those which undergo fast work. This, and the circumstance that shoeing generally removes the frog from contact with the ground, thus placing a heavier strain upon the flexor tendon of the foot as it passes over the navicular bone, is the chief cause in its production. It is rarely observed in the hind feet, because these have less weight to carry than the fore ones.

Symptoms.—The symptoms of this disease are generally well marked when it has advanced to a certain stage. In the stable the horse rests the affected foot by placing it forward and raising the heel slightly; if both feet are involved, then they are alternately rested; the foot is hot, and the litter in front is usually pawed behind. On coming out of the stable the horse steps short, and trotting on hard ground causes greater lameness than moving on soft soil. The horse is liable to trip and stumble, from the tendency to go on his toes, in order to relieve the back part of the foot. This tendency also causes the shoes to be worn more at the toes than elsewhere. If the disease is of some duration, the foot contracts, owing to the decrease in size of the frog, which leads to narrowing of the heels. Very often the frog is affected with thrush. The muscles of the limb, and especially those of the shoulder, become wasted, and the animal suffers much from the continual pain he experiences, and from which he can obtain no relief. This pain is increased if he be kept in a stall, and particularly if it has a sloping floor. When the disease is advanced it is almost impossible to mistake its existence.

Treatment.—The disease is, no doubt, largely preventable. In shoeing, the hoof should not be mutilated, but left strong and sound, and the frog ought to be intact and so prominent as to rest on the ground when the horse is standing. Horses should be kept in stalls which are level in front, or, better, in roomy loose boxes, and in the day time they ought to stand on moss litter. At the commencement of the disease the same procedure is to be observed, and exercise ought to be allowed on soft ground. Turning the horse out in a water meadow is most advisable. The toe of the hoofs should be kept short, and if shoes must be worn, and the wall is strong enough, the short imbedded or preplantar shoe ought to be preferred. If the horse is kept in the stable, cold water swabs must be applied to the feet, or the horse may stand in a foot bath of cold water for some hours every day. In some cases a cantharides blister applied to the coronets appears to be beneficial. In chronic cases, dividing the nerves of sensation that supply the foot (neurotomy) is the only palliative measure, and when the cases are judiciously selected, and the feet are carefully managed afterwards, such "unnerved" horses frequently perform good service.

Thoroughpin.

This is the name given to distension of the sheath of the tendon of the hind foot at the upper and back part of the hock. The tendon may be sprained or its sheath injured at this part, just in front of the point of the hock, and the swelling may be pushed from one side to the other—hence the name. The injury may occur in slipping or jumping, but the horses most liable to it are those with short hocks. There may or may not be lameness, but in either case the swelling is unsightly.

Treatment.—If the sprain is recent, then rest must be allowed, and a high-heeled shoe applied to the foot of the affected leg. Fomentations with warm water may be resorted to for some days, after which compound tincture of iodine should be painted over the swelling every day until the skin becomes slightly blistered; or the biniodide of mercury ointment may be applied two or three times at intervals of a week or ten days. When the swelling is chronic, then the pring truss, made for the purpose of applying pressure to his part of the hock, and sold by veterinary instrument makers, should be tried.

Bog Spavin.

This is not very common in light horses. It is a soft swelling on the front and inner part of the hock, above the seat of bone spavin, and is due to distension of the proper capsule of the hock joint. When the distension is great, there is also a swelling in the seat of thoroughpin, from the joint capsule being pushed upward and backward.

Bog spavin may appear without any assignable cause, but there has generally been a sprain, or series of sprains of the hock. Hunters in jumping, and stallions in covering are liable to this injury, especially if the hock is short.

Treatment.—This should be similar to that recommended for thoroughpin, the employment of the spring truss being even more beneficial for this condition than for the one just mentioned.

Wind Galls.

Wind galls are merely distensions of the sheaths of tendons below the knees and hocks, due either to rheumatism, sprain, or hard work—most frequently the last mentioned. They may, or may not be accompanied by lameness; if they are, then the soft puffy swellings are hot, and painful on pressure. They are most frequently seen about and immediately above the fetlocks of both fore and hind limbs.

Treatment.—If there is lameness, then the treatment should be as for sprain of the tendons; but if there is no lameness, but merely swelling, then equable pressure by means of bandages firmly and evenly applied, is the simplest and readiest treatment. Or if the horse can be rested for two or three months, a "charge" placed on the legs will have a good effect.

Cracked Heels.

Cracked heels are usually the result of leaving the skin in the hollow of the pasterns wet, especially in cold weather;

this is more likely to occur if the heels have been trimmed. The hind heels are most exposed to cracks or ulcerations. The lameness is more or less considerable, and especially when the horse first begins to move, when the pain may be so great as to cause him to travel for some distance on his toes. The injury is aggravated by sand and grit getting into the sores.

To prevent cracked heels, these should not be trimmed during cold wet weather, and when the horse returns to the stable they should be always carefully cleaned and dried.

Treatment.—When the skin is merely tender and inflamed, oxide of zinc ointment will be found soothing; or olive oil, four parts, and Goulard's extract one part, well mixed. If the heels are very painful they should be fomented with warm water and afterwards have a linseed meal and bran poultice applied to them for a day or two; after which the cracks may be dressed with carbolised oil, or with boric acid powder. When they are healed it is well to smear over the skin with a simple ointment composed of equal parts of beeswax and lard, to protect it from the weather for some time.

Mud Fever.

This is something akin to cracked heels, and generally arises from the same or similar causes. It is most frequently seen in hunters, and the skin over the body and limbs, the latter more particularly, is hot, painful, and rough; the animal moves with soreness, and the skin is tender to the touch, while the irritation may be so great as to excite some degree of fever, and even loss of appetite. The prevention of this disease is refraining from clipping the legs, and not washing them when the horse comes into his stable, but merely wisping off the superficial mud. When the legs are dry, then the dirt and dust may be brushed out. If the legs must be washed, this should be done with cold water and in a shed; then they should be thoroughly dried, hand-rubbed, and wrapped in woollen bandages.

Treatment.—If the limbs are very painful, it may be necessary to sponge them over with a solution of Goulard's extract of lead—one ounce to the pint of water ; or oxide of zinc ointment may be applied to the inflamed surface. A soothing liniment is composed of acetate of lead one ounce, olive oil one pint, water one pint ; to be well rubbed up, and applied by means of a piece of sponge. This liniment should be employed for some time after the skin has recovered, if the weather is cold and wet.

Surfeit.

This consists of an eruption on different parts of the body of small, hard lumps, generally accompanied with itching and symptoms of indigestion—the latter being probably the cause of the skin disturbance. It usually disappears in the course of two or three days, though sometimes it persists for two or three weeks.

Treatment.—A dose of laxative medicine, such as an ounce of powdered sulphur in a mash, which may be of bran or linseed, or a pint of linseed oil, with an ounce of nitrate of potass or carbonate of potass in the water given to drink once a day, will usually effect a cure. Care should be taken with the diet for some time.

Thrush.

Thrush is a diseased condition of the frog, generally due to its being cut away by the shoeing smith and thrown out of its function. It is marked by a foul-smelling discharge from the cleft of the frog, which becomes ragged and wastes, and may in time become tender and cause lameness.

Treatment.—It may be necessary to poultice the foot after all the loose portions of horn have been removed from the frog. The cleft should be thoroughly cleaned out by tow, then a little calomel ought to be pressed deeply into it, and maintained there by a pledget of tow. Subsequently, pledgets of tow smeared with Stockholm tar should be introduced into the cleft. If possible the frog should be allowed to come into contact with the ground, and the shoer ought to be prohibited from paring it.

Bone Spavin.

This is a bony enlargement at the inside and lower part of the hock. In some cases there is little if any enlargement, but two or more of the bones of the hock may be fixed together, or there may be ulceration between them. There is more or less stiffness or lameness in the joint, according to the extent and seat of the disease. The horse rests the leg very much, and goes somewhat on the toe of the foot. When he first begins to move the lameness is more marked than it is after he has travelled for some time. The lameness is sometimes very perceptible when the horse is moved in the stall. Spavin is most frequently seen in defectively shaped hocks, though it may occur from severe strain on well-shaped hocks, or working horses very hard when too young.

Treatment.—To be beneficial, treatment must be undertaken early. Absolute rest is indicated, and if the horse could be rendered immovable in the affected joint there would be a good chance of stopping the progress of spavin. But this is not possible, and all that can be done is to keep the horse quiet, a stall being preferable to a loose-box, and the animal can be tied up for some time. To ease the front of the joint, a high-heeled shoe may be placed on the foot, and either warm or cold fomentations applied to the hock for some days. Then biniodide of mercury ointment should be rubbed into the skin over the spavin, at intervals of a week or so. This treatment ought to be continued for six weeks or two months, when the result should be tested. If the lameness has not disappeared, then a seton should be passed over the seat of spavin, or firing may be resorted to, points being employed instead of lines.

Curb.

Curb is a sprain to the back part of the hock, at the upper part of the shank bone, and is manifested by a convexity or bulging, which is best seen when the hock is looked at sideways. It is often caused quite suddenly in jumping or slipping, and then there will probably be considerable lameness at first. Thin, short hocks, narrow at the bottom and somewhat angular, are most predisposed to curb.

Treatment.—When the sprain first occurs and there is lameness, the horse must be rested, a high-heeled shoe applied to the foot, and the hock fomented for a few days. Then a little biniodide of mercury ointment should be rubbed into the swelling, and repeated after an interval of a week. In about a fortnight the high-heeled shoe may be replaced by the ordinary one, and the horse allowed exercise every day.

Capped Hocks.

Capped hocks are the result of contusions, and rarely cause any amount of lameness, though they are unsightly.

Treatment.—If the injury is recent, and there is soreness and lameness, the contused hock should be fomented with warm water for two or three days, then treated with lead lotion. When the swelling becomes chronic, it may receive one or two applications of biniodide of mercury ointment; or a thick layer of soft pipeclay may be spread over it every day.

Injuries to the Foot.

The foot is more exposed to injury than any other part of the body. The most frequent injuries are treads, contusions, wounds from sharp objects while travelling on the road, pricks and bruises in shoeing, splitting of the hoof (sandcrack), bruise of the sole, corn, &c.

When the injury occurs to a part enclosed in the hoof, it is generally necessary to relieve the sensitive parts from pressure by removing the horn from over and around it, and preventing the shoe from touching it. When the inflammation runs high and there is much pain, fomentations and poultices are necessary, but they must not be continued for long; as a rule they should be succeeded by dry dressings. For injuries in which the hoof is involved, after the inflammation has been subdued, Stockholm tar is an excellent dressing, while it is a good protective.

Wounds.

Wounds are of different kinds, according to their mode of production—such as incised, punctured, contused, &c. The incised is that which is generally most easily repaired. When there is bleeding it should be checked as soon as possible by the application of cold or hot water, bandaging up the wound, applying pressure, or tying the bleeding vessel or vessels. Some chemical agents, such as perchloride of iron, are sometimes employed to check hæmorrhage.

If the wound is not large and the part can be bandaged, then after it is freed from dirt or other foreign matters, its edges should be brought together and the bandage applied, a piece of lint or fine tow being previously placed upon the wound. If it can be done, it is often advantageous to bring the sides of the wound together by means of one or more stitches of silk thread, or by brass pins passed through the skin on each side and a piece of tow or twine wound in figure of 8 fashion around the heads and points.

Bleeding from a punctured wound can generally be stopped by plugging it firmly with tow, lint, or any similar substance. The air should be excluded as early and as completely as possible from all wounds; so that after dirt or any other extraneous matter which may have gained access to them is removed, they should be carefully protected by tincture of myrrh, powdered boric acid, iodoform, or other antiseptic agent.

When the wounds are large and contused, it is generally not advisable, or possible even, to close them by sutures or close bandaging; as the dead portions have to be removed by the natural process of sloughing or suppuration. This process can often be expedited by fomentations with warm water.

Broken knees are a somewhat common accident, and the injury may vary from a slight skin graze to the most serious damage to all the soft tissues, and even the bones in front of the knee.

When such an accident happens, if the skin and other tissues are cut, the wound should be well cleaned by gentle washing with a sponge and warm or cold water; a piece of lint ought then to be placed over the part, and maintained there by means of a bandage. If the wound is not deep or very contused, I have seen some very good results obtained by dressing it, after it was well cleansed, with some tincture of myrrh and then applying a piece of lint on which Canada balsam was spread, over the injury, leaving it there until the place was healed. If the lint chanced to become detached, a fresh quantity of the balsam was spread over it and it was again stuck on.

When the wound is deep and contused and the joint probably opened, then after it has been freed from all foreign matter, the leg should be made immovable by means of a starched bandage or a long splint, or both combined, and extending from the hoof to above the knee; the portion of the bandage covering the wound being cut away, in order to permit the injury to be dressed. This dressing will depend upon circumstances, but I have found boric acid, or iodoform and starch powder in equal parts, and dusted over the surface, an excellent application. The horse should not be allowed to lie down during the treatment, and it is generally advisable to have him slung, so that he may rest his legs and not fall down until the wouud is healed.

Mange.

This is a rare disease in well-managed stables. It is very contagious, however, so that all horses are liable to become infected. It is caused by a microscopic insect, of which there are three kinds: one that infests the body more particularly, another that inhabits the neck and root of the mane and tail, and a third that confines itself usually to the legs. That which is far more frequently observed on light horses, is the one infesting the body. All cause intense itching, which impels the animals to bite and rub themselves almost continuously. The insect that burrows into the skin of the body, also produces shedding of the hair, and gives rise to the formation of crusts and raw places on the skin.

Mange is very contagious, and the parasites pass directly from affected to healthy horses, as well as through the medium of harness, clothing, litter, &c.

Treatment.—Cleanliness is a great obstacle to the extension of mange. Affected horses should be well washed with warm water and soft soap, applied by means of a scrubbing-brush; then when dry, the skin must be dressed with some agent that will kill the parasites. Before this is done, however, it is often found advantageous to soak the skin for some hours with a solution of carbonate of potass and oil. Afterwards an ointment composed of one part of tar oil and six parts of palm oil, will generally suffice to cure the disease; the ointment may be washed off in two or three days. In some obstinate cases the treatment may require to be repeated.

In addition to treating the animals, it is essential that clothing, harness, stable fittings, grooming tools, and everything else with which mangey horses may have been in contact, should be cleaned and dressed with a solution of carbolic acid, one part to five or ten of water.

Ringworm.

Ringworm is due to the presence of a microscopic vegetable parasite, which grows on the skin in such a manner as to produce more or less circular bare patches covered by a thin crust. It does not cause so much itching as the mange insect, though there is some; but it renders the skin unsightly, and may lead to considerable disfigurement if it is allowed to extend over the body. It most frequently affects young horses, and is very contagious.

Treatment.—This may be the same as that prescribed for mange, but it may be limited to the affected parts, and a little distance beyond them. An ointment composed of Stockholm tar one part, and lard three parts, answers very well.

Shoeing.

The management of horses' feet with the object of keeping them strong and healthy, is most important, and demands the constant attention of every horseman. The following rules should therefore be strictly observed, if horses are to be kept free from lameness and remain serviceable to a good old age, so far as shoeing is concerned :—

(1) Horses should be newly shod, or the old shoes removed and replaced, at least once a month.

(2) When being shod, the hoofs should be reduced to a proper length and evenly levelled, so that the toe will not be too long, nor one side higher than the other.

(3) The frog and sole should not be pared, interference with them being limited to removal of any loose portions.

(4) The shoe should not be heavier than is necessary to withstand wear for a certain period—say a month.

(5) The shoe should be made to fit the hoof—that is, be the full size of the circumference of the latter.

(6) The shoe should be attached to the hoof with as small and as few nails as may be necessary to keep it securely on the hoof.

(7) The nails should not be driven higher in the hoof than is needed to obtain a sound and firm hold.

(8) When the shoe is nailed on and the clenches laid down, the front of the wall should not be rasped, but left with its natural polish, and in all its strength.

(9) If possible, the frog should be allowed to come in contact with the ground.

INDEX.

Action of Cleveland Bays, 69
Action of Hackneys, 33, 38
American Trotting Horse, 94
American Trotting Register, 103
Antiquity of the Hackney, 27
Appetites of Ponies, 140
Arab Cross with Trotters, 84
Arabian Horse, 81
Arabs as Hunters and War Horses, 87
Asses, 153

Back Action of Hackneys, 40
Bellfounder (Jary's), 97
Bog Spavin, 213
Bone of Cleveland Bays, 69
Bone Spavin, 216
Breaking Hunters, 123
Breeding Hunters, 109
Breeding Hunters from Cleveland Bays, 74
Breeding from Roarers, 22
Breeding Trotters, 98
Breeds of Light Horses:
    Asses, 153
    Ponies, 136
    The American Trotting Horse, 94
    The Arabian Horse, 81
    The Cleveland Bay, 51
    The Hack, 127

Breeds of Light Horses—*contd.*
    The Hackney, 23
    The Harness Horse, 132
    The Hunter, 105
    The Thoroughbred, 1
    The Yorkshire Coach Horse, 76
British Horse, 2
Bronchitis, 198
Byerly Turk, 9

Cadet (Hackney), 49
Candidate (Hackney), 48
Capped Hocks, 217
Care of Young Foals, 180
Catarrh, 191
Chapman Horse, 51
Cleveland Bays, 51
Cleveland Bays as Agricultural Horses, 62
Cleveland Bays as General Utility Horses, 71
Cleveland Bay Horse Society, 68
Clothing Horses, 172
Colic, 204
Colour of Cleveland Bays, 54, 70
Condition of Hackneys, 43
Conformation of American Trotter, 101
Congestion of Lungs, 200
County Member (Hackney), 49
Cracked Heels, 213

INDEX.

Crossing with Cleveland Bays, 73
Curb, 217

Danegelt (Hackney), 49
Darley Arabian, 10
Decadence of Cleveland Bays about 1823, 63
Demand for Cleveland Bays in U.S.A., 66
Denmark (Hackney), 47
Diseases of Horses, 188
Draining of Stables, 162

East Anglia and Hackneys, 29
Eastern Blood in Hackneys, 26
Eastern Blood in Thoroughbreds, 3, 4
Eclipse, 11
Effect of Railways on Hackney Breeding, 34
Euren, H. F., on Hackney, 26, 31
Exercise, 173
Exmoor Ponies, 150

Families of Trotters, 97
Farcy, 196
Feeding Horses, 167, 173
Fever, 189
Fireaway (Triffit's), 26, 48
First Recorded Importations, 3
Flying Childers, 16
Foals, Care of, 180
Formation of Hunter Strains, 120

Ganymede (Hackney), 41
Gervase Markham, 5
Gilbey, Sir Walter, Bart., on Hunter Sires, 116
Glanders and Farcy, 196
Godolphin Arabian, 10
Grooming Horses, 171

Hackney Action, 38
Hackney Horse, 23
Hackney Horse Society, 23
Hackneys as Coach Horses, 42
Hackneys as Hunter Sires, 49
Hacks, 127
Hambletonian (American Trotter), 96
Handling Young Hunters, 122
Harness Horses, 132
Head of Hackney, 35
Height of Arabs, 83
Height of Hackneys, 31
Helmsley Turk, 6
Herod (Thoroughbred), 14
Hobbies, 7
Horses at Time of Roman Occupation, 2
Horses Imported from Turkey, Spain, and Naples, 3
Hunter Brood Mares, 110
Hunter Sires, 116
Hunters, 105

Importations, Early, 3
In-breeding of Ponies, 141
Inflammation of the Bowels, 205
Inflammation of the Lungs, 201
Influenza, 164
Initiating Young Horses into Harness Work, 126
Injuries to the Foot, 217
Injuries of Horses, 188
Irish Ponies, 151

Jumping, Teaching, 124

King Charles II. and Horse Breeding, 8, 12
Knee Action of Hackneys, 39

Lamenesses, 206

INDEX.

Laminitis, 203
Lawrence, John, on Hackneys, 27
London Hackney Shows, 24

Mambrino, 96
Management of Hackneys, 44
Management of Half-bred Stock, 122
Management of Light Horses, 159
Management of Trotters, 103
Mange, 220
Markham Arabian, 4
Masterman, T., and Cleveland Bays, 64
Messenger, 95
Modern Hackneys, 34
Mud Fever, 214
Mules, 156

Nancy Hanks (American Trotter), 98
Navicular Disease, 210
New Forest Ponies, 151

Original Shales, 26
Origin of Cleveland Bays, 52
Origin of Hackney, 23
Origin of Thoroughbred, 1
Osborne, Joseph, on Thoroughbreds, 10

Pacing Gait, 101
Park Hacks, 128
Paving of Stables, 162
Phenomena, 33
Pleurisy, 202
Points of American Trotters, 101
Points of Cleveland Bays, 69
Points of Hackney Brood Mare, 46
Points of Hackneys, 34
Points of Yorkshire Coach Horse, 79
Polo Ponies, 145

Ponies, 136
Pony Stud Books, 148

Queen's Premiums, 21

Reality (Hackney), 48
Redesdale, Lord, on Racehorses, 17
Revival of Cleveland Bay Breeding, 65
Rheumatism, 202
Ringworm, 221
Rous, Admiral, on Racehorses, 17, 18
Royal Mares, 9
Rufus (Hackney), 48

Saleable Horses, 114
Selection of Hackney Mares, 46
Shales, The Original, 26
Shetland Ponies, 151
Short Distance Races, 18
Shoeing, 221
Shoulders of Cleveland Bays, 73
Show Hunters, 115
Side Bone, 210
Size of Old Fashioned Hackney, 30
Soundness, 20
Speed of American Trotters, 98
Speed of Early Horses, 16
Splints, 209
Sprain of the Back Tendon, 207
Sprains, 206
Stable Fittings, 165
Stable Management, 167
Stables, 159
Standard Trotters, 104
Staying Powers of Thoroughbreds, 18
Strangles, 192
Suggested Re-Introduction of Arab Blood, 19

Sunol (American Trotter), 102
Surfeit, 215

Temper of Hackneys, 42
Thoroughbred Blood in Trotters, 100
Thoroughbred Crosses on Cleveland Bays, 59
Thoroughbred Horse, 1
Thoroughbreds as Hunters, 105
Thorough Pin, 212
Thrush, 215
Trapper, The, 139
Trotting of Hackneys, 33

Use of Ponies, 137

Ventilation of Stables, 160

Walking Action of Hackney, 41
Watering Horses, 170, 179
Weight Carrying Hunters, 112
Welsh Ponies, 152
What is a Pony? 143
William the Conqueror's Charger, 3
Wind Galls, 213
Working Cleveland Bays, 72
Worms, 206
Wounds, 218

Yorkshire Coach Horses, 76
Young Half-bred Stock, 122

www.ingramcontent.com/pod-product-compliance
Lightning Source LLC
Chambersburg PA
CBHW031341230426
43670CB00006B/401